Current Topics in Microbiology and Immunology

Volume 366

For further volumes:
http://www.springer.com/series/82

Current Topics in Microbiology and Immunology

Volume [illegible]

Series Editors

Klaus Aktories
Medizinische Fakultät, Institut für Experimentelle und Klinische Pharmakologie und Toxikologie, Abt. I, Albert-Ludwigs-Universität Freiburg, Albertstr. 25, 79104 Freiburg, Germany

Richard W. Compans
Department of Microbiology and Immunology, Emory University, 1518 Clifton Road, CNR 5005, Atlanta, GA 30322, USA

Max D. Cooper
Department of Pathology and Laboratory Medicine, Georgia Research Alliance, Emory University, 1462 Clifton Road, Atlanta, GA 30322, USA

Jorge E. Galan
Boyer Ctr. for Molecular Medicine, School of Medicine, Yale University, 295 Congress Avenue, room 343, New Haven, CT 06536-0812, USA

Yuri Y. Gleba
ICON Genetics AG, Biozentrum Halle, Weinbergweg 22, 06120 Halle, Germany

Tasuku Honjo
Department of Medical Chemistry, Faculty of Medicine, Kyoto University, Sakyo-ku, Yoshida, Kyoto 606-8501, Japan

Yoshihiro Kawaoka
School of Veterinary Medicine, University of Wisconsin-Madison, 2015 Linden Drive, Madison, WI 53706, USA

Bernard Malissen
Centre d'Immunologie de Marseille-Luminy, Parc Scientifique de Luminy, Case 906, 13288 Marseille Cedex 9, France

Fritz Melchers
Max Planck Institute for Infection Biology, Charitéplatz 1, 10117 Berlin, Germany

Michael B. A. Oldstone
Department of Neuropharmacology, Division of Virology, The Scripps Research Institute, 10550 N. Torrey Pines Road, La Jolla, CA 92037, USA

Rino Rappuoli
Novartis Vaccines, Via Fiorentina 1, Siena 53100, Italy

Peter K. Vogt
Department of Molecular and Experimental Medicine, The Scripps Research Institute, 10550 North Torrey Pines Road, BCC-239, La Jolla, CA 92037, USA

Honorary Editor: Hilary Koprowski
Biotechnology Foundation, Inc., Ardmore, PA, USA (deceased)

For further volumes:
http://www.springer.com/series/82

John S. Mackenzie · Martyn Jeggo
Peter Daszak · Juergen A. Richt
Editors

One Health: The Human–Animal–Environment Interfaces in Emerging Infectious Diseases

Food Safety and Security, and International and National Plans for Implementation of One Health Activities

Responsible Series Editor: Richard W. Compans

Springer

Editors
John S. Mackenzie
Faculty of Health Sciences
Curtin University
Perth, WA
Australia

Martyn Jeggo
CSIRO Australian Animal Health Laboratory
East Geelong, VIC
Australia

Peter Daszak
EcoHealth Alliance
New York, NY
USA

Juergen A. Richt
Department of Diagnostic Medicine/Pathobiology
Kansas State University College of Veterinary Medicine
Manhattan, KS
USA

ISSN 0070-217X
ISBN 978-3-642-35845-6 ISBN 978-3-642-35846-3 (eBook)
DOI 10.1007/978-3-642-35846-3
Springer Heidelberg New York Dordrecht London

Library of Congress Control Number: 2013934839

Printed on acid-free paper

Springer is part of Springer Science+Business Media (www.springer.com)

Preface

Global health security has become a major international concern. Our population faces imminent threats to human and animal health from the emergence and re-emergence of epidemic-prone infectious diseases, linked to the significant impact that these outbreaks are already having on national and international economies. The concept and drivers of disease emergence were clearly documented 20 years ago in the Institute of Medicine's seminal 1992 report, *Emerging Infections: Microbial Threats to Health in the United States*. (www.nap.edu/catalog.php?record_id=2008). This volume described the mechanisms leading to infectious disease emergence and highlighted possible strategies for recognizing and counteracting the threats. It has long been known that many of these diseases can cross the species barrier between humans, wildlife, and domestic animals; and indeed over 70 % of novel emerging infectious diseases are zoonotic, that is, they have their origins in animal reservoirs. There have been many examples of this since the Institute of Medicine's report two decades ago, including the emergence of H1N1 pandemic influenza virus, the SARS coronavirus, Nipah and Hendra viruses, Australian bat lyssavirus, Malaka virus, and avian influenza H5N1, to name but a few.

These diseases remind us that the health of humans, animals, and ecosystems are interconnected, and that to better understand and respond rapidly to zoonotic diseases at the human–animal–environment interfaces requires coordinated, collaborative, multidisciplinary, and cross-sectoral approaches. This holistic approach has been referred to as 'One Health', indicative of the commonality of human and animal medicine, and their connection to the environment. Although the concept is not new, 'One Health' has gained added momentum in the aftermath of the SARS epidemic of 2003 which posed the first major threat to human health and global economy of the new millennium. These concerns added to the mounting fears that highly pathogenic avian influenza H5N1 could develop into the next severe influenza pandemic. Not only would such a pandemic lead to significant mortality and morbidity, but the World Bank has estimated that it could cause a decline of up to 5% of global GDP (damages of US$3 trillion), causing far-reaching disruptions in the lives of people, communities, and countries. There are currently other potential threats from the new SARS-like coronavirus and H7N9 avian influenza virus. Thus there are compelling reasons to develop new approaches that will improve the detection, prevention, and control of zoonotic diseases.

In particular, it is essential that we breakdown the old concepts of professional silos and encourage a new era built around trust and multidisciplinary, cross-sectoral approaches.

The present momentum of 'One Health' can also be traced in part to the 2004 meeting of the Wildlife Conservation Society on 'One World, One Health: Building Inter-disciplinary Bridges to Health in a Globalized World'. The outcomes of the meeting were encapsulated in a series of 12 recommendations known as the Manhattan Principles that set priorities for an international, interdisciplinary strategy for combating threats to the health of life on Earth (www.oneworldonehealth.org). The momentum since 2004 has been maintained through a number of international ministerial meetings, including the International Ministerial Conferences on Avian and Pandemic Influenza (IMCAPI), which have been held to discuss issues relating to the spread, transmission, and possible containment of highly pathogenic avian influenza (H5N1), culminating at the 2010 meeting in Hanoi with the agreement between the Food and Agriculture Organization (FAO), the World Organization for Animal Health (OIE), and World Health Organization (WHO), entitled 'The FAO–OIE–WHO Collaboration: Sharing Responsibilities and Coordinating Global Activities at the Animal–Human–Ecosystems Interfaces'. The coordination between these three international organizations has also led to the formation of the Global Early Warning System for Major Animal Diseases including Zoonoses (GLEWS) which provides the intelligence essential to identify and ameliorate both human and animal diseases (www.glews.net) through sharing of information of disease events, epidemiological analyses, and risk assessments. In addition, it is highly probable that any new zoonotic disease would be detected through WHO's new International Health Regulations (2005) which are aimed at assisting countries in working together to save lives and livelihoods through a legal requirement for countries to rapidly detect and report outbreaks of disease of international concern.

This leadership is an essential component to operationalize 'One Health' ideals. Major scientific meetings have been held in Winnipeg through Health Canada and at Stone Mountain, Georgia through the Centers for Disease Control and Prevention, and by a wide variety of other interested groups such as the European Commission, joint meetings of FAO–OIE–WHO, Global Risk Forum (Davos), Institute of Medicine, the World Bank, APEC, and the Asian Development Bank. Many smaller, national, and regional meetings have also been held to further local 'One Health' planning. Of particular importance has been the information dissemination by the 'One Health' Initiative website (www.onehealthinitiative.com) and the more recently established 'One Health' Global Network's web portal (www.onehealthglobal.net) which have continued to build and sustain this momentum by providing a rapid means of communication and sharing data and news. As the field of 'One Health' matures, we have also begun to see the growing involvement of ecologists, wildlife biologists, environmental scientists, and the fusion of the fields of 'EcoHealth' and 'One Health'. There has also been considerable support for the 'One Health' approach in the United States through a partnership of major professional organizations that have formed the 'One Health' Commission, which brings together the American Medical Association, the American Veterinary Medical Association, the American Public Health Association, the Infectious Diseases Society of America, the Association of American Medical Colleges, and the Association of American Veterinary Medical Colleges. The inclusion of the latter two organizations is particularly relevant, breaking down professional barriers or silos through education. A number of universities and colleges are starting to respond with new 'One Health' courses; and one university, the University of Edinburgh, has developed a Masters postgraduate degree course.

More than 200 years ago, the German writer, artist, and politician, Johann Wolfgang von Goethe, reminded us that: "Knowing is not enough; we must apply. Willing is not enough; we must do." That epithet applies well to the 'One Health' movement, because in the midst of all the information that has been gathered about the health of humans, animals, and ecosystems, as well as the desire of many people in many nations and organizations to implement viable public health solutions, application, and action are essential. In this context, 'One Health' is not a new form of governance or a critique of existing patterns of governance. Rather, 'One Health' is a movement dedicated to building new levels of trust and transparency between disciplines, nations, organizations, and people. Such trust and transparency must begin with inspirational educational curriculums, teaching the next generation of clinicians and veterinarians how to apply and do their own work in such a way that many others come to appreciate the necessity of "One Health" in tackling difficult problems.

As these two volumes of Current Topics in Microbiology and Immunology go to press, many countries have established their own national 'One Health' policies and/or committees, recognizing the need to integrate and coordinate their human and animal surveillance to empower a more effective and rapid cross-sectoral response to zoonotic disease threats. There is little doubt that the 'One Health' concept will continue to develop and provide the coordinated, collaborative, multidisciplinary and cross-sectoral approaches essential to develop the rapid detection, and better predictive ability necessary for rapid response to future threats. In particular, we envisage a greater collaboration among environmental and ecological scientists with the animal and human health sectors of the 'One Health' movement. The linkages between the underlying socioeconomic and environmental drivers of emerging diseases, and the threat of pandemic emergence will likely be one area in particular where collaboration will be fruitful.

The purpose of these volumes is to present an overview of the 'One Health' movement, and in doing so, demonstrate the breadth and depth of its recent global development. The first volume has been divided into two Parts. The first part entitled "The Concept and Examples of a One Health Approach" examines 'One Health' from different perspectives especially that of human health and veterinary medicine, whether domestic or wildlife, the importance of understanding the different interfaces, the role of ecological science, and the compelling economics driving their cooperation and coordination. This is then followed by a series of examples where a 'One Health' approach has been useful in responding to specific diseases in the field. The second volume entitled "Food Safety and Security, and

International and National Plans for Implementation of One Health Activities" explores the importance of 'One Health' in food safety and food security. These are crucially important issues that are often not given the prominence they require and deserve as the world seeks to feed a growing population. This second volume also describes some of the international, regional, and national activities and plans to implement 'One Health' approaches. The final Part describes additional activities and approaches to strengthen the 'One Health' movement and increase its momentum in different ways. By reading, reflecting, and acting on the scale and depth of 'One Health' as set out in these volumes, you will be making your own contribution to the movement. Do not underestimate the importance of that contribution.

Contents

Contents of Volume 1: The Concept and Examples of a One Health Approach

Part II Examples of a One Health approach to specific diseases from the field

Contributors

Ochirpurev Ariuntuya WHO, P.O. Box 46/78, Ulaanbaatar, Mongolia, e-mail: ochirpureva@wpro.who.int; atuya_o@yahoo.com

Zayat Batsukh Government Implementation Agency for Veterinary and Animal Breeding, MoFA and Light Industry, Peace avenue-16a, SPB-IX, Ulaanbaatar, Mongolia, e-mail: zbatsukh@mail.mn

Barry Borman Centre for Public Health Research, College of Humanities and Social Sciences, Massey University, Private Box 756, Wellington, New Zealand

Naomi Cogger EpiCentre Institute of Veterinary, Animal and Biomedical Sciences, Massey University, Private Bag 11 222, Palmerston North, New Zealand

Ben Coghlan Burnet Institute, Melbourne, Australia, e-mail: coghlan@burnet.edu.au

Peter Collignon Infectious Diseases Unit and Microbiology Department, The Canberra Hospital, Canberra Clinical School, Australian National University, P.O. Box 11, Woden, ACT 2607, Australia, e-mail: peter.collignon@act.gov.au

Mathew Dixon Chatham House, London, UK

Adyadorj Dolgorkhand National Center for Infectious Diseases with Natural Foci, 20 Khoroo, Songinokhairkhan district, Central post office box 582, 18131 Ulaanbaatar, Mongolia, e-mail: khandadya@yahoo.com

Narelle Fegan CSIRO Animal, Food and Health Sciences, 671 Sneydes Rd, Werribee, VIC 3030, Australia, e-mail: Narelle.Fegan@csiro.au

Kari S. Gobius CSIRO Animal, Food and Health Sciences, 39 Kessels Rd, Coopers Plains, QLD 4108, Australia, e-mail: Kari.Gobius@csiro.au

Gyanendra Gongal Disease Surveillance and Epidemiology, WHO Regional Office for South East Asia, New Delhi 110 002, India, e-mail: Gongalg@SEARO.WHO.INT

Micah B. Hahn Nelson Institute, Center for Sustainability and the Global Environment (SAGE), University of Wisconsin, 1710 University Avenue, Madison, WI 53726, USA, e-mail: mbhahn@wisc.edu

David Hall University of Calgary, Calgary, Canada, e-mail: dchall@ucalgary.ca

Bernard Hang'ombe SACIDS at School of Veterinary Medicine, University of Zambia, Lusaka, Zambia

David L. Heymann Chatham House Centre on Global Health Security, London, UK; Infectious Disease Epidemiology, London School of Hygiene and Tropical Medicine, London, UK, e-mail: David.Heymann@hpa.org.uk

James Hope Animal Health and Veterinary Laboratories Agency Weybridge, Woodham Lane, New Haw, Addlestone KT15 3NB, Surrey, UK, e-mail: James.Hope@ahvla.gsi.gov.uk

Dominic Kambarage SACIDS at Sokoine University of Agriculture, Morogoro, Tanzania

Esron Karimuribo SACIDS at Sokoine University of Agriculture, Morogoro, Tanzania

Christopher Kasanga SACIDS at Sokoine University of Agriculture, Morogoro, Tanzania

Jean-Marie Kayembe SACIDS at School of Public Health, University of Kinshasa, Kinshasa, Democratic Republic of Congo

Kim Kayunze SACIDS at Muhimbili University for Health and Allied Sciences, Dar es Salaam, Tanzania

Juan Lubroth Animal Health Service, Food and Agriculture Organization, Viale delle Terme di Caracalla, 00153 Rome, Italy, e-mail: Juan.Lubroth@fao.org

Justin Masumu SACIDS at School of Public Health, University of Kinshasa, Kinshasa, Democratic Republic of Congo

Mecky Matee SACIDS at Muhimbili University for Health and Allied Sciences, Dar es Salaam, Tanzania

Joanna S. McKenzie EpiCentre Institute of Veterinary, Animal and Biomedical Sciences, Massey University, Private Bag 11 222, Palmerston North, New Zealand

Gerald Misinzo SACIDS at Sokoine University of Agriculture, Morogoro, Tanzania

Petra Muellner EpiCentre Institute of Veterinary, Animal and Biomedical Sciences, Massey University, Private Bag 11 222, Palmerston North, New Zealand

Aaron Mweene SACIDS at School of Veterinary Medicine, University of Zambia, Lusaka, Zambia

Luis Neves SACIDS at Faculty of Veterinary Medicine, Eduardo Mondlane University, Maputo, Mozambique

Dashdavaa Otgonbaatar National Centre for Zoonotic Diseases, Ministry of Health, 20 Khoroo, Songinokhairkhan district, 18131 Ulaanbaatar-211137, Mongolia, e-mail: da.otgon@yahoo.com

Jonathan A. Patz Nelson Institute, Center for Sustainability and the Global Environment (SAGE), University of Wisconsin, 1710 University Avenue, Madison, WI 53726, USA, e-mail: Patz@wisc.edu

Janusz T. Paweska SACIDS at National Institute for Communicable Diseases of the National Health Laboratory Services, Sandringham, Johannesburg, South Africa

Helen Ross The University of Queensland, St Lucia, Brisbane, QLD, Australia, e-mail: Helen.Ross@uq.edu.au

Carol S. Rubin National Center for Emerging and Zoonotic Infectious Diseases, Centers for Disease Control and Prevention, Atlanta, GA 30333, USA, e-mail: crubin@cdc.gov

Mark Rweyemamu SACIDS at Sokoine University of Agriculture, Morogoro, Tanzania, e-mail: mark.rweyemamu@btinternet.com

Jørgen Schlundt National Food Institute, Technical University of Denmark, Morkhoj Bygade 19, 2860 Soborg, Denmark, e-mail: jors@food.dtu.dk

Martin Simuunza SACIDS at School of Veterinary Medicine, University of Zambia, Lusaka, Zambia

B. Tsolmon Government Implementation Agency for Veterinary and Animal Breeding, MoFA and Light Industry, Peace avenue-16a, SPB-IX, Ulaanbaatar, Mongolia

Baatar Undraa National Center for Infectious Diseases with Natural Foci, 20 Khoroo, Songinokhairkhan district, Central post office box 582, 18131 Ulaanbaatar, Mongolia, e-mail: Undraa_56@yahoo.com

W. D. Vink EpiCentre Institute of Veterinary, Animal and Biomedical Sciences, Massey University, Private Bag 11 222, Palmerston North, New Zealand, e-mail: W.D.Vink@massey.ac.nz

Philemon Wambura SACIDS at Sokoine University of Agriculture, Morogoro, Tanzania

Peter R. Wielinga National Food Institute, Technical University of Denmark (DTU), Copenhagen, Denmark

Part I
Food Safety and Food Security: A One Health Paradigm

Food Safety: At the Center of a One Health Approach for Combating Zoonoses

Peter R. Wielinga and Jørgen Schlundt

Abstract Food Safety is at the center of One Health. Many, if not most, of all important zoonoses relate in some way to animals in the food production chain. Therefore, the food becomes an important vehicle for many, but not all, of these zoonotic pathogens. One of the major issues in food safety over the latest decennia has been the lack of cross-sectoral collaboration across the food production chain. Major food safety events have been significantly affected by the lack of collaboration between the animal health, the food control, and the human health sector. Examples range from BSE and *E. coli* outbreaks over dioxin crises to intentional melamine contamination. One Health formulates clearly both the need for and the benefit of cross-sectoral collaboration. In this chapter, we will focus on the human health risk related to zoonotic microorganisms present both in food animals and food from these animals, and typically transmitted to humans through food. We focus on these issues because they are very important in relation to the human disease burden, but also because this is the area where some experience of cross-sectoral collaboration already exist. Food related zoonoses can be separated in three major classes: parasites, bacteria, and viruses. While parasites often relate to very specific animal hosts and contribute significantly to the human disease burden, virus have often been related to major, well-published global outbreaks, e.g. SARS and avian- and swine-influenza. The bacterial zoonoses on the other hand often result in sporadic, but very wide-spread disease cases, resulting in a major

Prepared for: Current Topics in Microbiology and Immunology—'One Health' Special Edition.

P. R. Wielinga
National Food Institute, Technical University of Denmark (DTU), Copenhagen, Denmark

J. Schlundt (✉)
National Food Institute, Technical University of Denmark,
Morkhoj Bygade 19, 2860 Soborg, Denmark
e-mail: jors@food.dtu.dk

Current Topics in Microbiology and Immunology (2013) 366: 3–17 3
DOI: 10.1007/82_2012_238

Published Online: 5 July 2012

disease burden in all countries, e.g. *Salmonella* and *Campylobacter*. Next to these traditional zoonotic problems, the use of antimicrobials in (food) animals has also caused the emergence of antimicrobial resistant (AMR) zoonotic bacteria. It is important to realize the difference in the nature of disease epidemiology, as well as, in society's reaction to these diseases in different socio-economic settings. Some diseases have global epidemic—or pandemic—potential, resulting in dramatic action from international organizations and national agricultural—and health authorities in most countries, for instance as was the case with avian influenza. Other diseases relate to the industrialized food production chain and have been—in some settings—dealt with efficiently through farm-to-fork preventive action in the animal sector, e.g. *Salmonella*. Finally, an important group of zoonotic diseases are 'neglected diseases' in poor settings, while they have been basically eradicated in affluent economies through vaccination and culling policies in the animal sector, e.g. *Brucella*.

Contents

1 Introduction

How can food safety action play a key role in a generic One Health approach? Already in ancient times it was understood that humans could get sick from consumption of infected meat, and that keeping your animals healthy improved your own health. Our current health situation has much improved, simply by keeping animals healthy through good management and hygiene, vaccination programs and prudent drug treatment. Nevertheless, there are still many zoonotic diseases that threaten human health, including diseases hosted by all kinds of animals, ranging from wildlife to domestic animals, whether in companionship—or agricultural setting. Given the large amount and the obvious close contact we have with animals by raising/hunting, slaughtering, and eating them or their products (e.g. milk, eggs), food animals and wildlife form the largest reservoirs and production grounds for emerging zoonotic pathogens. Caused to some degree by our modern animal production methods, we now increasingly use antibiotics also in animal production, both for treatment and for growth promotion, thereby contributing significantly to the occurrence of an emerging risk: Anti Microbial Resistant (AMR) bacteria.

The action of authorities to protect society from zoonotic diseases differ significantly according to socio-economic status of the society in question, but also according to the zoonotic pathogen in question. Basically, zoonotic diseases related to food animals can be separated into three groups: Diseases with a potential for global spread with a dramatic public relation potential, such zoonotic diseases are often caused by virus and have resulted in dramatic action, including political action in most countries, e.g. SARS, avian influenza, and swine flu. Other diseases relate to the industrialized food production chain and are broadly distributed in all such production systems, which are found in all countries, rich or poor. The prevention of these diseases need to be considered along the full production chain, but most countries have yet to deal efficiently with these zoonoses, including pathogens such as *Salmonella* and *Campylobacter*. Finally, an important group of zoonotic diseases have been eradicated or drastically reduced in affluent economies through vaccination and culling policies or through introduction of hygienic management practices. In most poor settings these diseases are ‘neglected diseases’ which receive very little attention from national authorities or even international organizations. This group includes *Brucella*, bovine tuberculosis (TB), and cysticercosis to name a few.

Common to all groups is the potential for a dramatic reduction in the disease burden—as well as the economic repercussions—of these diseases through cross-sectoral surveillance and action. This means that a One Health approach represents a significant *potential for improvement.*

2 Transmission Routes

Through eating, direct contact, and via the environment, the human—and the animal bacterial flora are in contact with each other. Figure 1 outlines the most important transmission routes for bacteria between the human and the animal reservoir. Via these routes bacteria from (food) animals may enter the human reservoir and vice versa.

The foodborne route is probably the most important gateway for this contact. The vast majority of infections with enteric zoonotic bacterial pathogens, such as *Salmonella enterica*, *Campylobacter coli/jejuni*, and *Yersinia enterocolitica*, probably occurs through this route. For other zoonotic pathogens, direct contact

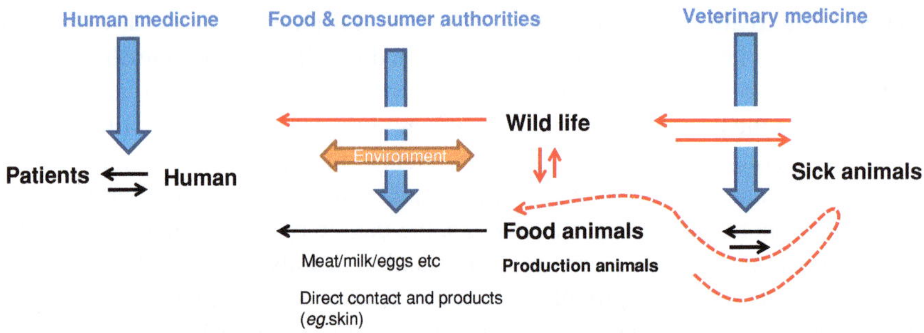

Fig. 1 Schematic presentation of important microbial transmission routes via which the human and (food) animals are in contact with each other. In *blue* control mechanisms are shown and in *red* some of the transmission routes that cannot be controlled, or escape control. Via the environment transmission may take place, through microbes present in excretion products and diseased animals, and in some countries also diseased humans. Next to the animals that are produced for direct consumption, in many developing countries animals are used as working animals to produce food and are thus included in this scheme. in these scheme. These animals when old or ill are often consumed (*red dashed line*), rather than destructed. Wildlife holds a broad spectrum of diseases including many highly pathogenic and deadly diseases. Though, the incidence may be low, because the associated risks are high, the consumption of wildlife animals and the spillover from wildlife to food/production animal is of importance. In Western countries wildlife is of lesser importance, however, in many developing countries and upcoming economies wildlife consumption may still be substantial, for instance in rural areas because of availability, as delicacies or other reasons. Moreover, in developing countries contact between humans and food animals may in general be easier and more frequent

between animals and humans may also be an important route of transmission, this could be the case for *Brucella* spp., *Enterohemorrhagic Escherichia coli* (EHEC) or some newer strains of Methicillin Resistant *Staphylococcus aureus* (MRSA). Bacteria from production animals are widespread in the environment, mainly as a result of their presence in manure. Thus, the environment and wild fauna also transforms into reservoirs of foodborne pathogens and resistance, and forms a source of (resistant) pathogenic bacteria into the food animals and human reservoirs. Although consumption of wildlife is not considered a major route in many developed countries, wildlife is consumed at a substantial level in developing countries. In addition, because of generally lower bio-safety levels in rural animal keeping, contact between humans and food animals may in general be more frequent in these countries. For instance, the general understanding now is that the SARS epidemic in 2003 originated in direct human contact with and/or consumption of, wildlife, or indirectly through contact between wildlife and domestic animals (Guan et al. 2003; Shi and Hu 2008). Wildlife holds a broad spectrum of diseases including some of the most deadly ones. For this reason also the consumption of wildlife animals, and the spillover of infectious diseases from wildlife to food/production animal, is of global importance.

3 Food Animal Zoonoses in General

Although a number of very important zoonoses are related to—and in some cases directly transmitted from wildlife animals—the vast majority of zoonotic disease cases in the world actually relate to animals bred for food purposes. Such zoonotic pathogens include bacteria, such as *Brucella*, *Salmonella*, *Campylobacter*, verotoxigenic *E. coli*, and *Leptospira*, parasites, such as *Taenia*, *Echinococcus*, and *Trichinella* or virus, such as Influenza A H5N1 (Avian influenza) and Rift Valley Fever virus. It of course also includes 'unconventional agents', such as prions, of which the most well-known is the one causing Bovine Spongiform Encephalopathy (BSE) in cows and new variant Creutzfeldt-Jakob disease in humans.

Diseases originating on the farm can in many cases be efficiently dealt with on the farm. For example, brucellosis in animals (mainly cattle and sheep) has been eliminated in many countries, thereby virtually eliminating the human disease burden (Godfroid and Käsbohrer 2002). Also, some of the main parasites can be effectively controlled at the farm level, and this could work for both *Taenia solium* in pigs (defined by WHO/FAO/OIE as a 'potentially eradicable parasite'), as well as, *Trichinella spiralis* (in many animals, including pigs); both have essentially been eliminated from farmed pigs in several northern European countries (WHO/FAO/OIE 2005; Gottstein et al. 2009).

3.1 Zoonoses Related to the Food Production Chain

Outbreaks of viral diseases in humans, originating in or spreading through farm animals (avian flu—H5N1 and 'swine flu'—H1N1) have caused major global alerts in the last decade. These zoonotic, global influenza outbreaks (H1N1 even characterized by WHO as a pandemic) spread very quickly either in the animals (H5N1) or directly in the human population (H1N1). Although the total human disease burden related to the endemic bacterial zoonoses are probably manifold higher than these influenza outbreaks, it is basically these relatively few (but clearly global) outbreaks that have put One Health on the global agenda. The failure to predict or even monitor disease spread in animals in order to link this to the prevention of disease in humans, presented regulators, and politician with a wake-up call regarding the need for cross-sectoral collaboration between the animal and human health sectors.

In contrast to the dramatic viral outbreaks, bacterial food-related zoonoses are usually occurring endemically in farm animals. These pathogens are found in most food animals produced in industrialized settings. It should be realized that most countries—including most developing countries—have a significant part of the food animal production in some sort of industrialized setting. Such settings are invariably linked to a number of important zoonotic pathogens, including *Salmonella*, *Campylobacter*, and *Escherichia coli*. These pathogens, while

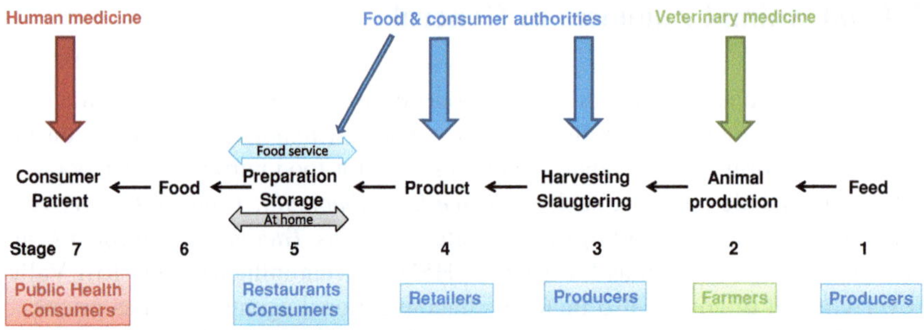

Fig. 2 Farm-to-fork scheme showing how infectious diseases may travel through the food chain. We have arbitrary defined seven stages that may be distinguished in the production chain of most animal derived foods. For individual food types, or non animals derived foods, different chains may be drawn in a similar way. Different controlling organizations are presented in the top of the picture and different stakeholders are presented in the lower part of the figure

widespread and endemic and in reality causing a major global disease burden, are not often recognized as important human pathogens. Part of the reason for this is the absence of a One Health framework, a framework that could ensure cross-sectoral collaboration and data-sharing and thereby lay the foundation for a realistic description of the situation, as well as, of potential sensible solutions. There are some countries (especially in northern Europe) that have instituted cross-sectoral data collection for zoonoses, typically through a construction called Zoonosis Centers. The data sharing across animal, food, and human health sectors have enabled science-based solutions, resulting most noticeably in significant reduction in human salmonellosis through lowering *Salmonella* prevalence in animals (Wegener et al. 2003). These constructions and solutions are clearly following One Health principles, and have basically done so since 1994! Similar solutions would be relevant in all countries with industrialized food animal production, but it is noteworthy that efficient solutions in this (sometimes called commercial) part of a national production system often will have repercussions also down to the traditional poor farmer (sometimes referred to as the communal production system). For instance, *Salmonella enteritidis* entered Zimbabwe through imported animals (poultry) in the commercial sector around 1993, and thereafter spread quickly to the communal sector, as well as, to the human population (Matope et al. 1998). The background for this is most likely that old animals from the commercial sector are sold to communal production systems. And where the animal goes the pathogen goes—therefore lowering the prevalence in the commercial sector would enable a reduction in prevalence also in the communal sector, thereby in turn lowering the human incidence of disease.

The spread of foodborne zoonoses through the food production chain has for more than 20 years been referred to as a 'farm-to-fork' (or 'boat-to-throat') issue related to the different stages of food production, but often originating at the farm (or all the way back to the feed used at the farm). This realization clearly represents original One Health thinking, and it should be noted that risk mitigation solutions under this framework typically have focused on a consideration of the full food production continuum, involving all relevant stakeholders. Figure 2 tries to capture a generalized picture of such a chain, starting with animal feed and ending in human consumption of animal-based food products.

3.2 Zoonoses Related to Poverty

Whereas zoonotic diseases with pandemic potential, such as avian or 'swine' influenza and SARS, have received committed attention from world leaders, and while zoonotic diseases related to industrialized food production systems have received some recognition leading to—at least in some countries—efficient risk mitigation action, a number of very important zoonotic diseases, disproportionately affecting poor and marginalized populations, are largely ignored.

These types of zoonoses are many, and the prevalence in animal populations vary according to local agricultural, demographic, and geographic conditions. For many such diseases solutions to dramatically decrease the disease burden are well-known, but action is lagging (for example, for many of the parasitic zoonoses). WHO refers to such diseases as 'Neglected Diseases' (Molyneux et al. 2011).

The group of Neglected Zoonoses include bacterial diseases, such as brucellosis (with significant sequelae), leptospirosis, Q-fever (with high mortality), and bovine TB. Bovine TB appears to be increasing in many poor settings with HIV infections as an important factor for progression of TB infection to active TB disease. For both, brucellosis and bovine TB the disease in cattle causes lowered productivity, but seldom death, and both infections have been largely eradicated from the bovine population in the developed world, by a test and slaughter program, thereby in effect eliminating this human health problem (Godfroid and Käsbohrer 2002).

Important zoonotic parasitic diseases include schistosomiasis, cysticercosis, trematodiasis, and echinococcosis, several of which with significant mortality rates or long-term sequelae including cancer and neurological disorders. Cysticercosis is emerging as a serious public health and agricultural problem in poor (García et al. 2003). Humans acquire *Taenia solium* tapeworms when eating raw or undercooked pork meat contaminated with cysticerci. The route of transmission is, pigs are infected through *Taenia* eggs shed in human feces, and the disease is thus strongly associated with pigs raised under poor hygienic conditions. This again means that the cycle of infection can be relatively easily broken when introducing efficient management, as has been the case in most developed countries.

Given that 70 % of the rural population in poor countries is dependent on livestock and working animals to survive (FAO 2002), the effect of these animals carrying a zoonotic disease can be dramatic, both relative to human health directly, but also as it affects the potential to earn an income. This also affects the potential mitigation action; for instance the large-scale culling of animals, which can be a viable solution in rich countries, might be problematic in the poorest countries. Such solutions would mean not only loss of food, but also a serious socio-economical disruption, in some cases leading to national instability.

4 AMR in Food Animals

In the early 1940s, antibiotics were first introduced to control bacterial infections in humans. The success in humans led to their introduction in veterinary medicine in the 1950s, where they were used in both production and companion animals. Next to agricultural animals, antibiotics, nowadays, have also found their way into intensive fish farming and some are used to control diseases in plants. Their use is thus wide-spread.

Antibiotics in animals are used essentially in three ways: for therapy of individual cases, for disease prevention (prophylaxis) treating groups of animals, and as antibiotic growth promoters (AGP). For AGP use, antibiotics are added to animal feed at sub-therapeutic concentrations to improve growth. The mechanism by which this works was (and still is) unclear, nevertheless, this type of antibiotic use led to a steep increase in antibiotic consumption when it was introduced. In general, when first introduced, the use of antibiotics led to improved animal health and most likely to higher levels of both food safety and food security. The use of antibiotics therefore sky-rocketed. Between 1951 and 1978, the use in the United States alone went from 110 to 5580 tons (WHO 2011).

However, the use of antibiotics in animals has over the years also resulted in a selective pressure for AMR microorganisms, contributing significantly to the human health problem of AMR bacteria; notably a number of bacterial strains that were previously susceptible to antibiotics are now, in very high frequencies, becoming resistant to these antibiotics, some of them representing very important or even last resort treatment potential for humans (Bonten et al. 2001). Nowadays, there are serious efforts by national authorities and some international organizations to reduce the antibiotic overuse in animal production (WHO 2011; FAO/OIE/WHO 2003), especially—but not only—through abolishing their use as AGP. However, there seems to be major problems in ensuring cross-sectoral understanding, and indeed cross-sectoral solutions in this area. The veterinary profession and the medical profession is seen as accusing each other of AMR problems, and in a sense they are both right—all use of antimicrobials can cause AMR, therefore both animal and human use cause problems. But in order to achieve a science-based understanding of the problem, we need data on both animal and human uses, and about both AMR in bacteria in animals, in food, and in humans. Therefore, a One Health approach in which all stakeholders work together will be necessary to investigate the problems and provide science-based solutions that can efficiently reduce the spread of AMR bacteria from animals to humans (most often through food) and vice versa (most often through the environment).

5 National and International One Health Efforts to Control Food Related zoonoses

5.1 Efforts to Contain Zoonoses

Clearly, through the increase in global trade and travel 'the world has become a village'. Food we eat today could have come from animals raised yesterday, thousands of miles away. To combat zoonotic foodborne zoonoses we need improvements and adjustments in our food production systems based on a global vision and approach. Internationally, different organizations have recognized that combating zoonoses is best done via a One Health approach. The World Health Organization (WHO), the World Organization for Animal Health (OIE), and the Food and Agriculture Organization of the United Nations (FAO), agreed a seminal paper: 'A Tripartite Concept Note' (FAO/OIE/WHO 2011) in which they express the need to collaborate for a common vision. Given the impact zoonotic disease have in particular on the most vulnerable sectors in our societies, the World Bank and the United Nations Children's Fund (UNICEF), as well as, the United Nations System Influenza Coordinator (UNSIC), share this One Health strategy to combat zoonotic disease (UN 2008). A One Health approach will open solution scenarios that are now not considered for treatment of the zoonoses problems because of the costs involved. For instance, while vaccination in some cases is the ultimate tool to prevent disease, it is often not considered because the costs of mass vaccination can be much higher than the public health benefit savings. Getting a true picture of the cost for the different stakeholders and setting up a framework for sharing of estimates of cost, as well as, mitigation strategies could likely enable (new) ways of reaching sensible solutions (Narrod et al. 2012).

While collaboration and control at the international level can help prevent the global spread of zoonotic disease and facilitate outbreak control at national levels, this is not enough. Many countries have established specialized infectious disease- or zoonosis centers in which zoonotic disease in particular are the focus of the work, and which help to establish and coordinate collaboration between the different sectors. Many of these work centers and these initiatives do focus on zoonoses originating from food animals, both from animals kept in industrialized settings and wildlife animals.

To monitor zoonotic outbreaks, many national authorities and relevant experts at country level, report important outbreaks that have the potential of cross-border spread to WHO, under the auspices of the International Health Regulations (WHO 2005). In addition, reporting is also often done on a more voluntary basis to ProMED-mail (http://www.promedmail.org), which is an internet-based Program for Monitoring Emerging Diseases worldwide, setup by International Society for Infectious Diseases and supported by many different (anonymous) institutes and individuals. The program is dedicated to rapid global dissemination of information on outbreaks of infectious diseases and acute exposures to toxins that affect human health, including those in animals and in plants grown for food or animal feed, and thereby supports the One Health principles.

Many of the (international) organizations and governing bodies named above have generated guidelines to control—and disseminate information about–food related zoonoses, such as for instance, WHO's Global Foodborne Infections Network (GFN) (www.who.int/gfn), the European Food Safety Authorities (EFSA) (www.efsa.europa.eu/en/topics/topic/zoonoticdiseases), Foodnet from the US Center for disease control (www.cdc.gov/foodnet), the Med-Vet-Net Assosiation (www.medvetnet.org)and others. The goal of these networks is essentially the same: *To help capacity-building and promote integrated, laboratory-based surveillance and intersectional collaboration among human health, veterinary and food-related disciplines to reduce the risk of foodborne infections.*

5.2 Efforts to Contain AMR Zoonoses

The emergence of AMR in (food) animals is now getting significant attention, and One Health principles have been suggested to deal efficiently with these problems. Collaboration between the FAO/WHO Codex Alimentarius Commission and the OIE (the World Animal Health Organization) have generated important guidance, on how an integrated approach and the prudent use of antimicrobials can help to reduce the risk of the emergence of AMR in (food) animals, and thereby in humans. This guidance for the prudent use of antibiotics can be found on their respective website (e.g. www.codexalimentarius.org; www.who.int/foodborne_disease/resistance; http://www.oie.int/our-scientific-expertise/veterinary-products/antimicrobials/).

Already before this One Health approach was initiated 'Global Principles for the Containment of Antimicrobial Resistance in Animals Intended for Food' have been generated (WHO 2000). The three major tenets of these principles are:

- Use of antimicrobials for prevention of disease can only be justified where it can be shown that a particular disease is present on the premises or is likely to occur. The routine prophylactic use of antimicrobials should never be a substitute for good animal health management.
- Prophylactic use of antimicrobials in control programs should be regularly assessed for effectiveness and whether use can be reduced or stopped. Efforts to prevent disease should continuously be in place aiming at reducing the need for the prophylactic use of antimicrobials.
- Use of antimicrobial growth promoters that belong to classes of antimicrobial agents used (or submitted for approval) in humans and animals should be terminated or rapidly phased-out in the absence of risk-based evaluations.

These Global Principles have been supplemented with, (1) guidance on the prudent use of antibiotics from the Codex Alimentarius Commission together with OIE, and (2) six priority recommendations from WHO to reduce the overuse/

misuse of antibiotics in food animals for the protection of human health (WHO 2001), which are:

1. Require obligatory prescriptions for all antibiotics used for disease control in (food) animals.
2. In the absence of a public health safety evaluation, terminate or rapidly phase out the use of antibiotics for growth promotion if they are also used for treatment of humans.
3. Create national systems to monitor antibiotic use in food-animals.
4. Introduce pre-licensing safety evaluation of antibiotics (intended for use in food animals) with consideration of potential resistance to human drugs.
5. Monitor resistance to identify emerging health problems and take timely corrective actions to protect human health.
6. Develop guidelines for veterinarians to reduce overuse and misuse of antibiotics in food animals.

In the latest publication of the WHO (regional office for Europe (WHO 2011)), which covers the broader scope of AMR in relation to both animal and human use, a One Health approach is promoted to help tackle the rise of AMR and seven recommendations have been suggested: To strengthen national multisectoral coordination for the containment of antibiotic resistance; to strengthen national surveillance of antibiotic resistance; to promote national strategies for the rational use of antibiotics and strengthen national surveillance of antibiotic consumption; to strengthen infection control and surveillance of antibiotic resistance in health care settings; to prevent and control the development and spread of antibiotic resistance in the food-chain; to promote innovation and research on new drugs and technology; and to improve awareness, patient safety, and partnership.

Many countries have generated programs to contain zoonoses and AMR zoonoses based on these UN Principles and Guidelines. In particular, the Danish program to contain AMR zoonoses has gained international attention and has been analyzed in different publications (WHO 2003; Aarestrup et al. 2010; Hammerum et al. 2007). The reason for this may have been the early One Health approach which the Danish government and different stakeholders proposed to combat AMR. The decision to do so came after publication of the finding that 80 % of *Enterococci* in chicken were highly resistant to vancomycin, which is a last resort drug for human therapy (Wegener et al. 1999). The reaction to this dramatic finding was the initiation of a knowledge— and collaboration platform to combat AMR: the Danish Integrated Antimicrobial Resistance Monitoring and Research Program (DANMAP) in 1995, supported by the Danish Ministries of Health, of Food, Agriculture and Fisheries and of Science, Technology, and Innovation. Figure 3 shows how DANMAP is organized and how the three sectors (animal health, food safety, and public health) work together.

The objectives formulated for DANMAP, and which have been updated over time, are: (1) to monitor the consumption of antimicrobials used in (food) animals and humans, (2) to monitor the occurrence of AMR in (zoonotic) bacteria in animals, food, and humans, (3) to study the associations between antimicrobial consumption and antimicrobial resistance, and (4) to identify routes of

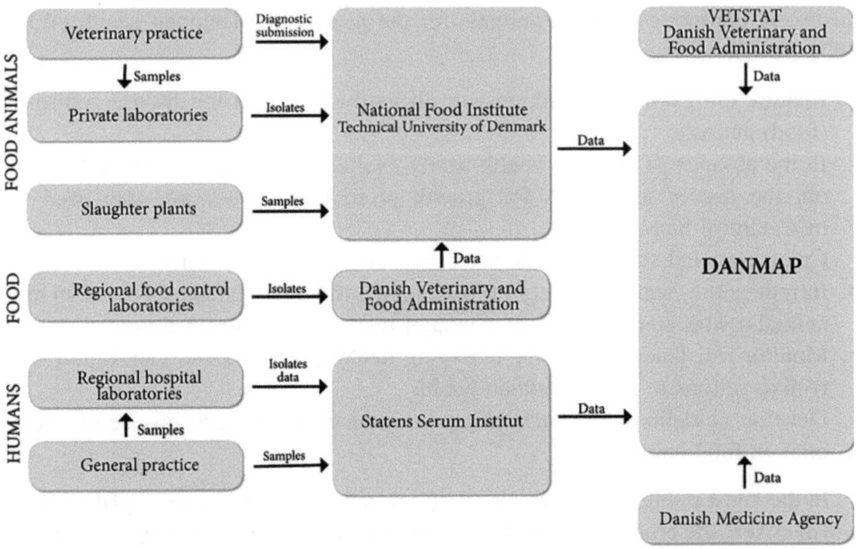

Fig. 3 Organization of DANMAP showing how the different institutes and agencies work together and how the information on AMR in humans, animals, and food is brought together in DANMAP (taken from: www.danmap.org)

transmission and areas for further research studies. One of the findings from DANMAP was that data on drug usage are essential for a good understanding of the problem. In Denmark, an automated program called Vetstat has been introduced to collect quantitative data on all prescribed medicine for animals from veterinarians, pharmacies, and feed mills (Stege et al. 2003). With this information, it has been possible for the Danish Veterinary and Food Authority (DVFA) to introduce "The Yellow Card Initiative" (DVFA 2012). Like in a football match, farmers and veterinarians get a card when their antimicrobial use is excessive, and only by reducing the antibiotic use (for instance, by adopting management methods from low users) they can lose the card. This does not only work as a stick, it also gives the users a sense of how well they are doing compared to their colleagues. In the European Union, several countries now have started to collect similar data to compare antibiotic use at country level (EMA 2011).

6 A Global Identifier for One Health Microbiology

To understand and anticipate the transmission of foodborne infections, and infectious diseases in general, it is important to monitor infectious diseases at critical points throughout the feed-to-food chain. To do this efficiently, different sectors should collect data in a harmonized way, so data can be compared and integrated. Historically, however, different sectors have been using different techniques. Since the publication of the first sequenced human genome in 2003, DNA sequencing has taken a giant leap forward. For public health as well as for veterinary science, whole genome sequencing (WGS) may take disease diagnostics to a new level. The potential efficiency has already been shown by the tracking of the massive cholera outbreak in Haiti in 2010 and more recently, in diagnosing the multi-resistant *Klebsiella* Oxa48 outbreak in a hospital in Rotterdam, the Netherlands (Potron et al. 2011). And in 2011, the serious EHEC outbreak in Germany was traced back to Egyptian imported fenugreek seed using WGS (Mellmann et al. 2011).

In Brussels, September 2011, an international group of scientists from all over the world, with representatives from OIE, WHO, EC, USFDA, US CDC, E-CDC, universities, and public health institutes concluded that, although the technology to do WGS for microorganisms is available, a global genomic database to make efficient use of WGS information is still missing (Kupferschmidt 2011). Such a database is needed and should be open to, and supported by, scientists from all fields: human health, animal health, and food safety, and should include genomic data for all types of microorganisms as well as meta-data to trace back the source of the microorganism. Building such a database is only possible through a One Health approach on all levels. There should be cooperation internationally, across sectors (e.g. human, animal, and food), as well as, between different stakeholders. Persuing such a major goal will not only be beneficial for the developed world, but maybe especially for developing countries. For them, genomic identification will be a giant leap forward in the fight against infectious diseases, and could be likened to the spread of cell phones, which made expensive and exclusive land-lines unnecessary and made communication possible for everybody. Identification and typing of microorganisms will suddenly become technically and economically feasible, enabling control and prevention efforts previously missing in many regions. At the same time, developing countries moving to use this technology will not need to develop expensive specialized lab systems since microbiological lab work will basically be the same for bacteria, parasites, and viruses. If set up in a sensible, inclusive, open-source framework WGS analysis will provide the world with a strong weapon in the fight to combat infectious diseases in all sectors.

7 Conclusions

A One health approach may be synergistic in controlling zoonotic diseases to support both sufficient food safety and sustainable food security. Clearly, because of the unique situation of transmissibility between humans and animals, zoonosis control relies on the control of the microorganisms in (1) animals, (2) the food chain, and (3) humans. In addition, as many zoonoses find their origin in animals before being transmitted to humans, the most effective intervention is often achieved at the source, i.e. the animals or, when this is not possible, by blocking the transmission to humans.

The approaches that need to be taken to reduce the risk of human disease from food animal zoonoses should include involvement of all stakeholders from the human and animal health side. They should work together to keep and improve animal health and animal food production such that potentially harmful zoonoses get the lowest chance of surviving in animals or entering the transmission route to humans. At the transmission level, it will be of major importance to involve food and consumers authorities and related stakeholders to make sure the spillover from the animal reservoir is kept as low as possible.

The level of involvement of the different stakeholders will differ per country or type of disease. Given that 70 % of the rural population in poor countries is dependent on livestock and working animals to survive, the effect of these animals carrying a zoonosis will work out differently than in the industrialized countries. It is important to realize that the focus on zoonotic pathogens with a potential for dramatic global spread (such as avian influenza) is significantly higher than the focus on endemic zoonotic pathogens in the food production chain, even though such pathogens are globally distributed and cause a dramatic disease burden (e.g. *Salmonella*). And it is as important to realize that a number of the most important zoonoses relate directly to food production systems in poor settings, that could be reduced dramatically through well-known interventions (such as *Brucella*, bovine TB, and cysticercosis).

References

Aarestrup FM et al (2010) Changes in the use of antimicrobials and the effects on productivity of swine farms in Denmark. Am J Vet Res 71(7):726–733

Bonten MJ, Willems R, Weinstein RA (2001) Vancomycin-resistant *enterococci*: why are they here, and where do they come from? Lancet Infect Dis 1:314–325

Danish Veterinary and Food Administration (DVFA) (2012) The yellow card initiative. http://www.foedevarestyrelsen.dk/english/Animal/AnimalHealth/Pages/The-Yellow-Card-Initiative-on-Antibiotics.aspx

EMA (2011) Trends in the sales of veterinary antimicrobial agents in nine European countries (2005–2009). European Medicines Agency (EMA/238630/2011). http://www.ema.europa.eu/docs/en_GB/document_library/Report/2011/09/WC500112309.pdf

FAO/OIE/WHO (2003) Joint FAO/OIE/WHO expert workshop on non-human antimicrobial usage and antimicrobial resistance: scientific assessment, Geneva, 1–5 Dec 2003. http://www.who.int/foodsafety/publications/micro/en/amr.pdf

FAO/OIE/WHO (2011) The FAO-OIE-WHO collaboration—a tripartite concept note. http://www.fao.org/docrep/012/ak736e/ak736e00.pdf

Food and Agricultural Organisation of the United Nations (FAO) (2002) Improved animal health for poverty reduction and sustainable livelihoods. FAO animal production and health paper 153. http://www.fao.org/DOCREP/005/Y3542E/Y3542E00.HTM#Contents

García HH, Gonzalez AE, Evans CA, Gilman RH (2003) Cysticercosis working group in Peru *Taenia solium* cysticercosis. Lancet 362(9383):547–556. PubMed PMID: 12932389

Godfroid J, Käsbohrer A (2002) Brucellosis in the European Union and Norway at the turn of the twenty-first century. Vet Microbiol 90(1–4):135–145

Gottstein B, Pozio E, Nöckler K (2009) Epidemiology, diagnosis, treatment, and control of trichinellosis. Clin Microbiol Rev 22(1):127–145. PubMed PMID: 19136437

Guan Y, Zheng BJ, He YQ, Liu XL, Zhuang ZX, Cheung CL, Luo SW, Li PH, Zhang LJ, Guan YJ, Butt KM, Wong KL, Chan KW, Lim W, Shortridge KF, Yuen KY, Peiris JS, Poon LL

(2003) Isolation and characterization of viruses related to the SARS coronavirus from animals in southern China. Science 302(5643):276–278

Hammerum AM et al (2007) Danish integrated antimicrobial resistance monitoring and research program. Emerg Infect Dis 13(11):1632–1639

Kupferschmidt K (2011) Epidemiology. Outbreak detectives embrace the genome era. Science 333(6051):1818–1819. PubMed PMID: 21960605

Matope G, Schlundt J, Makaya PV, Aabo S, Baggesen DL (1998) *Salmonella enteritidis* in poultry: an emerging zoonosis in Zimbabwe. Zimbawe Veterinary J 29:132–138

Mellmann A, Harmsen D, Cummings CA, Zentz EB, Leopold SR, Rico A, Prior K, Szczepanowski R, Ji Y, Zhang W, McLaughlin SF, Henkhaus JK, Leopold B,BielaszewskaM, Prager R, Brzoska PM, Moore RL, Guenther S, Rothberg JM, Karch H (2011) Prospective genomic characterization of the German enterohemorrhagic *Escherichia coli* O104:H4 outbreak by rapid next generation sequencing technology. PLoS One 6(7):e22751. Epub 20 Jul 2011. PubMed PMID: 21799941

Molyneux D, Hallaj Z, Keusch GT, McManus DP, Ngowi H, Cleaveland S, Ramos-Jimenez P, Gotuzzo E, Kar K, Sanchez A, Garba A, Carabin H, Bassili A, Chaignat CL, Meslin FX, Abushama HM, Willingham AL, Kioy D (2011) Zoonoses and marginalised infectious diseases of poverty: where do we stand? Parasit Vectors 14(4):106

Narrod C, Zinsstag J, Tiongco M (2012) A one health framework for estimating the economic costs of zoonotic diseases on society. EcoHealth. doi:10.1007/s10393-012-0747-9

Potron A, Kalpoe J, Poirel L, Nordmann P (2011) European dissemination of a singleOXA-48-producing *Klebsiella pneumoniae* clone. Clin Microbiol Infect 17(12):E24–26. PubMed PMID: 21973185

Shi Z, Hu Z (2008) A review of studies on animal reservoirs of the SARS coronavirus. Virus Res 133(1):74–87

Stege H, Bager F, Jacobsen E, Thougaard A (2003) VETSTAT-the Danish system for surveillance of the veterinary use of drugs for production animals. Prev Vet Med 57:105–115

UN (2008) Contributing to one world, one health; a strategic framework for reducing risks of infectious diseases at the animal–human–ecosystems interface. WB, WHO, UNICEF, OIE, FAO and UNSIC

Wegener HC, Aarestrup FM, Jensen LB, Hammerum AM, Bager F (1999) Use of antimicrobial growth promoters in food animals and *Enterococcusfaecium*resistance to therapeutic antimicrobial drugs in Europe. Emerg Infect Dis 5(3):329–335. PubMed PMID: 10341169

Wegener HC, Hald T, Lo Fo Wong D, Madsen M, Korsgaard H, Bager F, Gerner-Smidt P, Mølbak K (2003) Salmonella control programs in Denmark. Emerg Infect Dis 9(7):774–780. PubMed PMID: 12890316

WHO (2000) WHO global principles for the containment of antimicrobial resistance in animals intended for food: report of a WHO consultation with the participation of the food and agriculture organization of the united nations and the office international des Epizooties, Geneva, Switzerland, 5–9 June 2000 World Health Organization, Geneva. http://whqlibdoc.who.int/hq/2000/WHO_CDS_CSR_APH_2000.4.pdf

WHO (2001) WHO global strategy for containment of antimicrobial resistance. http://www.who.int/drugresistance/WHO Global Strategy English.pdf

WHO (2003) Impacts of antimicrobial growth promoter termination in Denmark. The WHO international review panel's evaluation of the termination of the use of antimicrobial growth promoters in Denmark. World Health Organization, Geneva. http://www.who.int/gfn/en/Expertsreportgrowthpromoterdenmark.pdf

WHO (2005) International health regulations (2005), 2nd edn. http://www.who.int/ihr/9789241596664/en/index.html

WHO (2011) Tackling antibiotic resistance from a food safety perspective in Europe. http://www.euro.who.int/__data/assets/pdf_file/0005/136454/e94889.pdf

WHO/FAO/OIE (2005) Guidelines for the surveillance, prevention and control of taeniosis/cysticercosis. http://www.oie.int/doc/ged/D11245.PDF

The Importance of a One Health Approach to Preventing the Development and Spread of Antibiotic Resistance

Peter Collignon

Abstract Antibiotic resistance is a continuing and growing problem. Antibiotic resistance causes increased deaths, complications, expenses and prolonged hospital stays. There are not likely to be many new classes of antibiotics becoming available in the next few decades. We need to take a "One Health" perspective to this problem. We need to preserve the usefulness of those antibiotics we currently have by decreasing their overall use in all sectors, and especially the use of broad spectrum agents. We also need to improve our ability to prevent infections and the spread of resistant bacteria wherever they arise or are found. This means improving our practices with infection control, hygiene and animal husbandry. We need to improve the development and the delivery of effective and safe vaccines to prevent infections. We need safe water supplies. Our failure to do this has already resulted in large numbers of people entering a "post-antibiotic era" for many common infections.

Contents

P. Collignon (✉)
Infectious Diseases Unit and Microbiology Department,
The Canberra Hospital, Canberra Clinical School,
Australian National University, P.O. Box 11,
Woden ACT 2607, Australia
e-mail: peter.collignon@act.gov.au

Current Topics in Microbiology and Immunology (2013) 366: 19–36
DOI: 10.1007/82_2012_224

Published Online: 13 June 2012

1 Introduction

The ever increasing problem of antimicrobial resistance demonstrates why the concept of "One Health" is so vital. This concept helps us to better understand why the problem of antimicrobial resistance is currently so pervasive and then to understand how we should best intervene to improve the situation.

Antibiotic resistance is growing at an ever increasing rate in almost all bacterial species that cause disease in people and animals. There is also little prospect that any new class of effective drugs will be developed and become available to use in the next 10 years to treat serious infections caused by these resistant bacteria. We are now seeing more and more people die with common infections that were easily treatable with effective antibiotics 20 years ago. This is a growing pandemic (Carlet et al. 2011).

For many people around the world, particularly those in developing countries, they have already returned to the preantibiotic era because of rising resistance in some of the most common bacteria causing serious bacterial infections in people, e.g. *Escherichia coli*. An ever increasing proportion of these infections are now untreatable. For many infections caused by *E. coli* in countries such as India and China, there are no longer any readily available and effective oral therapies. Even when people are able to afford and access hospital care, no effective injectable antibiotics may be available. Unfortunately, there continues to be dramatic rises and spread of resistant bacteria—more recently Gram-negative strains resistant to broad spectrum cephalosporins and carbapenems (Kumarasamy et al. 2010; Walsh et al. 2011).

Antimicrobials are essential drugs. They are needed to maintain the health and welfare of people. In people, serious bacterial infections remain common and include bloodstream infections, meningitis, pneumonia and peritonitis. In the era before antibiotics, blood stream infections with *Staphylococcus aureus* and *Streptococcus pneumoniae* were associated with mortality rates of over 80 % (Finland et al. 1959). Antimicrobials are also important for animal health. Antimicrobial drugs are not effective against resistant bacteria.

The antibiotics last century was correctly hailed as "miracle drugs". Their development and use quickly led to dramatic decreases in the mortality and morbidity rates of common life-threatening bacterial infections. These dramatic observed affects, however, had a major downside—the continuing massive overuse of antibiotics in both people and in food animals. This overuse was also of poor efficacy. In people, most antibiotic use is mainly for conditions where the benefits are either nonexistent or marginal (e.g. viral infections, bronchitis, etc.). However, the majority of antibiotic volumes used in the world continue to be used in food animals. In contrast to people this antimicrobial use is not for the therapeutic use of individual sick animals. The vast majority of use of antimicrobials in food animals (when usage can be measured) continues to be for growth promotion and for mass prophylaxis.

Antibiotic resistance is harmful and associated with worse outcomes. Serious infections caused by resistant bacteria do not respond as well to therapy and are associated with higher mortalities and prolonged hospital stays (Carlet et al. 2011; Cosgrove et al. 2003; de Kraker et al. 2011; Klevens et al. 2007; Tumbarello et al. 2010; Wang and Chen 2005).

1.1 What are the Consequences of Antimicrobial Resistance?

For people infected with resistant bacteria there are many additional problems;

- Antibiotics need to be used are often much more expensive (e.g. linezolid compared to ampicillin to treat enterococcal infections).
- Antibiotics that need to be given intravenously instead of orally (e.g. for *E. coli*, meropenem instead of oral amoxycillin).
- Antibiotics with lower intrinsic activity need to be used (e.g. vancomycin compared to flucloxacillin to treat resistant *S. aureus* infections).
- There may be in the future no antibiotic available for them that is active against the bacteria causing their infection.

These factors result in

- increased deaths,
- increased complications,
- additional expenses,
- prolonged hospital stays,
- additional toxicity and
- the need to for intravenous therapy rather than oral therapy.

In food animals the consequences are similar (although the examples will be different).

1.2 One Health, the Environment and Antimicrobial Resistance

Wherever antibiotics are used, resistant bacteria eventually develop and spread. This occurs both with people and animals. These resistant bacteria spread from person to person, animals to animals, people to animals and animals to people. They contaminate waterways when excretions or waste from either people or animals enters these waterways. They also are found frequently in food produced from animals that have received antibiotics. Slaughter processes and distribution networks result in cross contamination of many food products with resistant bacteria.

The extensive use of antimicrobials in all sectors (human and agriculture) means these drugs are often also found in the environment, especially in waterways and soils where bacteria are then again exposed to these drugs, often in low concentrations. Antibiotics are commonly used in aquaculture and in horticulture (e.g. gentamicin and streptomycin to spray apples). This results in much easier ingress of residual antibiotics used into waterways (Diwan et al. 2010; Mayerhofer et al. 2009; Zhou et al. 2011).

People and animals often come in contact with or ingest resistant bacteria because exposure to them is so widespread. If water and/or foods have antibiotic residues in them, then people and animals also ingest these residues. Acquired resistant bacteria are frequently carried by people and animals. These bacteria are then often reexposed to more antibiotics. A positive and very harmful feedback loop then results that leads to very high rates of resistant bacteria being found in many people and animals (or their products e.g. foods).

All sectors (people, animals and the environment) are directly and indirectly interconnected. A "One Health" approach to the problem of antimicrobial resistance means interventions will be better targeted against the complete and intertwined picture and not just, as occurs now, at subsections of this total picture.

1.3 Antibiotic Resistance is an Inevitable Consequence of Use

Genes encoding for antibiotic resistant are present naturally in the environment for most, if not all, antibiotics. This is because most antibiotics are derived from precursors that are "natural products" produced by fungi or higher order bacteria to help them survive against other microbial competitors (Davies and Davies 2010; Webb and Davies 1993). The organisms' that produce these antibiotics, however, usually need mechanisms to protect themselves against the effects of these toxic products they produce (after all antibiotics are designed to kill microorganisms). This means that often the microorganisms that produce antibiotics also have resistance genes and antibacterial products such as beta-lactamases etc. (Davies and Davies 2010; Webb and Davies 1993). This means whenever antibiotics are widely used, it is almost inevitable that resistance will develop as bacteria acquire resistance genes already present in the wider environment and these bacteria will then have a competitive advantage.

The greater the quantity of antibiotics used the more resistance will eventually develop. Resistant bacteria rapidly and easily move from site to site and from country to country. Because so much of this resistance is encoded by mobile genetic elements, the genes can move into other bacteria including quite different species. Thus the essential element to control antimicrobial resistance is to limit and decrease the amounts of antibiotics used in all sectors (i.e. human, agriculture, the environment). We need also to keep people and animals healthy so they do not need to receive antibiotics because they have less illness (good animal husbandry, immunisation etc.) and to stop the spread of resistant bacteria by better hygiene and infection control practices.

1.4 Resistant Bacteria and Genes Spread Easily

Resistant bacteria move readily from person to person, from hospital to hospital, from food animals to people and from country to country (Aarestrup et al. 2008b; Collignon et al. 2009; FAO 2003; Huijsdens et al. 2006; JETACAR 1999; Kennedy and Collignon 2010). Many spread through water and in foods. Foods (especially meats) also frequently contain bacteria that are multiresistant. Water is commonly contaminated with bacteria. When water is heavily contaminated with either human or animal faecal waste, then multiresistant bacteria can persist and even be distributed via chlorinated water supplies (e.g. in New Delhi with the MBL *E. coli* strains) (Walsh et al. 2011).

The bacterial genes that encode for antimicrobial resistance can also be readily transmitted between bacteria of the same species and also between bacteria of different species (Carlet et al. 2011; Kumarasamy et al. 2010; Walsh et al. 2011).

1.5 We Can Decrease the Spread of Resistant Bacteria and Prevent Them Causing Disease

Infection control interventions help. In the UK a national program decreased the number of MRSA bacteraemia episodes per year from 2003 to 2007 by over 40 % (from 3,955 to 2,376 episodes) (Health Protection Agency 2009).

Immunisations such as Hib were very effective in decreasing the amount of resistant Hib that was seen and was causing increasing problems 20 years ago (Collignon et al. 2008a, b). This is an example where immunisation has had a profound effect on decreasing the number of resistant bacteria causing disease. Similar effects have also been seen with Pneumococcus with a successful conjugated vaccine with much less disease now in children and therefore less antibiotics having to be used to treat children because they have less serious disease from this organism (Collignon et al. 2008a, b). In animals and fish, vaccines have been very successful at preventing disease and through decreasing antimicrobial use (e.g. salmon and fluoroquinolone use in Norway) (Markestad and grave 1997).

Clean water is an essential component in controlling antimicrobial resistance. Water is likely to be the major vehicle, particularly in the developing world, where resistant bacteria spread from person to person. This means keeping contaminated animal waste and human waste out of water ways as much as is practicable and ensuring water is treated to such a standard that it minimises the risk that pathogens and commensal bacteria carrying resistant genes will be ingested by people or animals. Clean water considerably decreases the amount of GIT disease and transmission of *Salmonella*, *Campylobacter* and many other pathogens. This means less illness and thus less need for antibiotics.

We need to prevent multiresistant bacteria from being in our food supply. The best way to achieve this is to stop the use of "critically important" antibiotics in our food animals and better limit the use of all antibiotics in food animals. We can also decrease the number of organisms in food by better controls along the food chain such as the way animals are slaughtered so that there is less contamination of the carcass with bowel bacteria and find other ways to decrease the number of resistant bacteria in foods. At the other end of the food chain, after the food is produced, issues such as pasteurisation of milk and eggs or other heat treatment can considerably decrease the number of pathogens and therefore antibiotic resistant bacteria that are in the food that is subsequently distributed to consume. Obviously consumer education can also help stop cross contamination from uncooked foods and cooked foods but also to foods such as lettuce, tomatoes etc., which may not be cooked before they are ingested.

The use of animal and human manure to grow food is also an issue if it contains large numbers of pathogens including resistant bacteria that may not be inactivated or removed before the product reaches the market. This can have implication in the globalisation of food trade. A recent example was the Haemorrhagic *E. coli* outbreak in Germany. The bean sprouts involved came from bean seeds imported from Egypt. The seeds are presumed to have been contaminated by human or animal waste in Egypt. When germinated and grown in Germany, the *E. coli* already present, increased markedly in numbers. Then because these sprouts were not cooked before being ingested, large numbers of people became ill. This resulted in huge pressures on the hospitals and Intensive Care systems in Germany and was also associated with many deaths (CDC 2011).

1.6 Resistance to Antimicrobials Classified as "Critically Important" is Common

Increasing resistance is occurring in almost all medically important bacteria, including to antimicrobials classified as "critically important" or "last line" for human health (Collignon et al. 2009). This means when resistance is present there will be very limited or no antimicrobial therapy that will still be effective to treat infections caused by these resistant bacteria. In hospitals we see increasing numbers of bacterial infections to which there are no effective antibiotics available. This includes infections caused by *E. coli*, *Acinetobacter* spp., *Serratia* spp. and *Enterobacter* spp. (Carlet et al. 2011; Fernando et al. 2010; Li et al. 2006; Walsh et al. 2011).

Resistance rates in almost all types of bacteria are much higher in developing countries. For most people living in developing countries this problem is compounded by poor access to appropriate diagnostic facilities, less resources being available to help institute and maintain appropriate hygiene and infections control practises as well as difficulties in accessing adequate and affordable medical care.

Many classes of critically important antibiotics are also used in food production animals. The most important of these from a human health consequence perspective have been identified by the World Health Organisation (WHO) as fluoroquinolones, third- and fourth-generation cephalosporins and macrolides (Collignon et al. 2009; WHO 2009).

1.7 The Drug Pipeline is Empty

Most antimicrobial classes were discovered decades ago. There have been very few new classes of antibiotics developed in the last 30 years (fluoroquinolones, lipopeptides, oxazolidinones). There have been some developments in classes of antibiotics that have already existed that have led to agents with much improved activity (ketolides and tigecycline). However, these latter two agents are just variations of macrolides and tetracycline's respectively (Carlet et al. 2011; Collignon et al. 2009).

The problem we have is that antibiotic resistance is developing much faster than there are any new drugs or drug classes likely to be available in the near future. This is particularly a problem for Gram-negative bacteria where there does not even look that there are many promising drugs in any advanced research stage yet alone in the development pipeline. The financial rewards for pharmaceutical companies to research and then market completely new classes of antibiotics is relatively poor compared to the returns on drugs that need to be taken by a large percentage of the population continuously e.g. cholesterol-lowering drugs (Collignon et al. 2008a, b; Power 2006). Unfortunately this situation is not likely to change in the near future.

1.8 Surveillance is Essential

We need much better and timely surveillance of antimicrobial usage and of resistant bacteria—locally, nationally and internationally. The results need to be readily available so we can better see what is happening with resistance in different areas and needs to involve both the human and non-human sector. We need to know the volumes and types of antimicrobials being used. This will allow not only the better choice of empiric antibiotic therapy but also help us better target problem areas with preventive interventions, improved antibiotic stewardship and other programs. This will then help to stop or slow resistance from getting worse in those targeted areas and hopefully even reverse some of the resistance levels seen.

2 What are the Most Important Bacteria We Need to Worry About?

Almost all bacteria that cause infections in people have higher rates of antimicrobial resistance present now compared to 10 or 20 years ago. Some infections, however, are more common and cause more serious infections in people. *E. coli*, *S. aureus*, *Enterococcus* spp. and *S. pneumoniae* are the most common bacteria causing serious infections in people (Beidenbach et al. 2004; Collignon et al. 2005; Collignon et al. 2011; Decousser et al. 2003; ECDC 2010; HPA 2009; Kennedy et al. 2008).

The more important examples of human infections are discussed below.

2.1 *Escherichia coli*

Escherichia coli is the commonest cause of serious bacterial infections in people. In the developed countries, bloodstream infections occur at rates between 30 and 60 episodes per 100,000 people each year (ECDC 2010; Kennedy et al. 2008) and are associated with substantial mortality and morbidity. There are likely over 2 million bloodstream infections per year worldwide. *E. coli* causes substantially more infections but which are not usually life-threatening e.g. urinary tract infections.

We are seeing rapidly increasing rates of antimicrobial resistance, including multiresistance. In many developing countries antimicrobial resistance is extensive and widespread and few or no agents may be available for therapy. Intravenous carbapenems e.g. meropenem can usually still be used to treat most infections. But even to these agents, resistance appears to be rapidly developing. These agents are usually only available in intravenous forms and are expensive. This effectively means that many people cannot access any antibiotics that are effective for these very common infections (Carlet et al. 2011).

The main reservoir for *E. coli* is the bowel and there is a large turnover every day of *E. coli* (Collignon and Angulo 2006; Corpet 1988; Johnson et al. 2006). While many *E. coli* strains are relatively specific in where they both live and multiply (e.g. some may be adapted for the pig gut), a large proportion of *E. coli* carried by people are acquired via foods and especially from poultry (Johnson et al. 2006). This is particularly the case for antibiotic resistant bacteria (Johnson et al. 2006). In many developed countries *E. coli* remains largely sensitive to third-generation cephalosporins, fluoroquinolones and/or aminoglycosides and these agents can usually still be used to treat those with serious infections. However, this is not the case in countries especially developing countries (Walsh et al. 2011). Travellers from countries with low rates of resistance to critically important antimicrobials such as fluoroquinolones and third-generation cephalosporins, often acquire these bacteria when visiting countries with much higher endemic rates of infections—most likely via food and/or water. Carriage of these resistant bacteria can be over 50 % in travellers and persist after returning home for over 6 months (Kennedy and Collignon 2010; Tängdén et al. 2010).

We are seeing increasing levels of ESBL *E. coli* in developed countries including the US and Europe. These strains are resistant to all third- and fourth-generation cephalosporins, are often community acquired and foods are a source. In particular poultry has been found to be frequently contaminated with multiresistant *E. coli* (Brinas et al. 2003). In Hong Kong, ESBL rates in poultry *E. coli* isolates were 78 % (Ho et al. 2011). In people there is now a worldwide epidemic with resistant *E. coli* carrying encoding CTX-M and CMY ß-lactamases (Aarestrup et al. 2008a, b; Cavaco et al. 2008; Mesa et al. 2006; Zhao et al. 2001). In Europe there are 100,000's episodes per year and blood isolates in 2009 showed ranges in different countries of 4–29 % for ESBL's and 9–44 % for fluoroquinolone resistance. It is of note that ESBL bacteraemia in Europe is associated with a mortality of 32 % within 30 days of their sepsis (de Kraker et al. 2011).

2.2 Staphylococcus aureus

Staphylococcus aureus is commonly carried asymptomatically by people in the community, particularly in their noses and on skin. It is also present in many food animals such as poultry and pigs. In people it is one of the most common, virulent bacteria that cause infections especially healthcare associated infections (Beidenbach et al. 2004; Collignon et al. 2008a, b; Collignon et al. 2005; ECDC 2010; HPA 2009). Even now, when we have good medical support for patients in hospitals (including intensive care), if a person has *S. aureus* bacteraemia then their median mortality rate is 25 %. If they have an antibiotic resistant variety (e.g. MRSA), then their then with bacteraemia their median mortality rate is 35 % (Cosgrove et al. 2003).

Serious infections are very common. In Denmark the annual rate of all *S. aureus* bloodstream infections is about 28 per 100,000 inhabitants per year. In the USA *S. aureus* bloodstream infection rates may be as high as 50 per 100,000 per year (or about 150,000 episodes per year) (Collignon et al. 2005; ECDC 2010). In Australia it is about 30 per 100,000 populations (Collignon et al. 2005).

Rates of the more resistant varieties of *S. aureus* (i.e. MRSA) are very high. In the USA and in many European countries as many as half of all *S. aureus* isolates causing bloodstream infections are MRSA (ECDC 2010; Klevens et al. 2007). In hospital the percentage caused by MRSA are even higher. In the US it is estimated that there may be over 100,000 episodes of invasive MRSA infections per year, mainly bacteraemia (Klevens et al. 2007).

Recent developments have resulted in more agents that are effective against *S. aureus* and other Gram-positive bacteria becoming available. This has included newer antibiotics such as linezolid, tigecycline and daptomycin. However, resistance, associated toxicity and/or high cost have limited their use. These agents also do not appear to be more effective than vancomycin. Vancomycin is less active than beta-lactam antibiotics against methicillin sensitive strains of *S. aureus* (MSSA). Thus this also means that one other clinical cost of increasing resistance is the need to use drugs that are intrinsically less active in serious disease (Collignon et al. 2008a, b).

The increasing numbers of community MRSA strains that are not healthcare related is a major concern. These are now causing a large and increasing percentage of community acquired infections in the US, Europe and Australia and elsewhere. In some cities over 50 % of community *S. aureus* infections are now MRSA. This means that for very common infections, we now need to use antibiotics that are more expensive, more toxic and less effective than agents we could previously depend on.

MRSA strains also develop and spread in food animals. Similar factors drive this development and spread as happens in people—over use of antibiotic especially broad spectrum agents, crowding and poor housing, social disadvantage and less than optimal infections control and/or hygiene. Recently MRSA strains have been found that spread from pigs to human (e.g. in the Netherlands and Denmark) and cause infections in people (Aarestrup et al. 2008a, b; Khanna et al. 2008; Lewis et al. 2008).

2.3 *Streptococcus pneumoniae*

Streptococcus pneumoniae is spread from person to person. It is a common cause of pneumonia, meningitis, otitis media and bloodstream infections (Collignon and Turnidge 2000; Hsueh et al. 1999; Pallares et al. 1995). It does not have non-human reservoirs and thus all the resistance we likely results from antibiotic use in people and/or associated poor hygiene (that allows the spread of this bacteria from person to person).

Increasing levels of resistance are seen to all antibiotics, particularly to penicillins. One antibiotic that can still be relied on in all circumstances to treat serious pneumococcal disease (including meningitis) is vancomycin although its penetration into CSF is relatively poor and it is not absorbed when given orally. Other agents such as linezolid appear to be effective as resistance in pneumococcus is currently very low. Oral therapy is very important for the treatment of many infections other than meningitis. High dose oral amoxicillin appears to be effective when therapy is needed even if intermediate penicillin resistance is present. However, with other oral agents, unfortunately increasing numbers of pneumococci are developing resistance to tetracyclines, co-trimoxazole and macrolides, which limits therapeutic options such as the oral treatment of pneumonia and other conditions (Collignon and Turnidge 2000; Hsueh et al. 1999; Pallares et al. 1995).

2.4 Other Gram-Negative Bacilli

There are many Gram-negative bacteria that cause serious disease particularly in health care settings (Collignon et al. 2008a, b). Examples include *Enterobacter* spp., *Pseudomonas aeruginosa, Serratia, Klebsiella* and *Acinetobacter*. Some may be untreatable with any antibiotic including polymixins (Hujer et al. 2006); Fernando et al. 2010). Other examples include *P. aeruginosa* and *Burkeholderia* spp. where now frequently there are no effective antibiotics that can be used in those with serious infections such as acquired in intensive care units, patients with cystic fibrosis and complicated lung infections. An older and relatively toxic antibiotic (polymixin) is increasingly being used as IV therapy as no other option may often be available to treat these resistant bacteria (Li et al. 2006). Fosfomycin is also being increasing used to treat multiresistant Gram-negative infections.

2.5 Enterococcus

Enterococcus species in particular *Enterococcus faecium* are intrinsically resistant to large numbers of antimicrobials. In people most infections are caused by *Enterococcus faecalis* which remains normally sensitive to both ampicillin and vancomycin (Collignon et al. 2008a, b; Heuer et al. 2006; Moellering 2005). For some serious infections such as endocarditis, an aminoglycoside needs to be added to ampicillin to achieve bactericidal activity. If high-level resistance to aminoglycosides is present, then endocarditis (which in the preantibiotic era had 100 % mortality) will not usually be able to be cured.

Enterococci are intrinsically resistant to cephalosporins. This is likely an important reason why they are selected out and are increasing in numbers in environments such as hospitals where cephalosporins are frequently used. Enterococci are becoming increasingly important pathogens in hospital and cause many serious infections such as bloodstream infections. Of particular concern are vancomycin resistant Enterococci (VRE) as there are only limited options for therapy and it readily spreads within a hospital environment given its hardiness both to environmental stressors and also to disinfectants and cleaning. There is also the concern that the genes that encode for vancomycin resistance may spread to more virulent bacteria such as *S. aureus*. Fortunately compared to 10 years ago we now have more agents available to treat infections caused by VRE (e.g. linezolid).

In most hospitals infection control practices try to limit the spread of these bacteria by isolating patients and requiring increased precautions to be taken by all medical and nursing staff looking after them e.g. gowns, gloves and isolation rooms. Thus the appearance of VRE is a concern particularly if also found in foods as was the case when avoparcin was used extensively as a growth promoter in the past. Other resistant enterococci can also spread via foods to people (Aarestrup et al. 2008a, b; Heuer et al. 2006).

2.6 Food-borne Pathogens (Salmonella and Campylobacter)

Antimicrobial resistance is increasing in many food-borne pathogens in particular *Salmonella* and *Campylobacter.* Agents that were very effective in the past including ciprofloxacin are now ineffective. (Aarestrup et al. 2008a, b; Engberg et al. 2001; Iovine and Blaser 2004; Mead et al. 1999; Pegues et al. 2005; Unicomb et al. 2003). This fluoroquinolone resistance is clearly related to fluoroquinolone use in food animals.

Infections with non-typhi *Salmonella* strains are common in developed countries (and even more common in developing countries). In developed countries nearly all these strains are derived from food animals. Increasing antibiotic resistance is an issue in these bacteria as well and some have been impossible or very difficult to treat. Of particular concern is the development of ESBL as when this occurs there may be no therapy available to treat pregnant women or children if they develop serious infections (e.g. bacteraemia) as third-generation cephalosporins are the drug of choice in that circumstance. ESBL *Salmonella* strain can develop from the use of third-generation cephalosporins in poultry. In Canada a close association has been found between the use of third-generation cephalosporin (ceftiofur) ESBL *Salmonella* and ESBL *E. coli* (CIPARS 2007).

Salmonella typhi is a pathogen that spreads from person to person usually via contaminated food and water. It has no animal reservoir and thus all resistance is likely the result of what antibiotics given to people along with poor hygiene and poor water infrastructure. If improved water supply and sewage infrastructure were introduced this would also have a significant effect in decreasing numbers of these infections (including antibiotic resistant infections).

Campylobacter is the commonest cause of bacterial diarrhoea in developed countries. The main causative organism is *Campylobacter jejuni* and mainly derived from poultry as its initial source. Increasing resistance is seen in these strains to both fluoroquinolones and macrolides. For most cases, no antibiotic therapy is needed. However, with more severe disease, fluoroquinolones and macrolides are the agents of choice and thus this resistance is problematic. Wherever fluoroquinolones have been used in poultry resistance develops and spreads and can reach very high rates in countries such as in Spain (Aarestrup et al. 2008a, b; Collignon et al. 2008a, b). Even in the US, where only a small percentage of poultry were exposed to fluoroquinolones, ciprofloxacin resistance rates in *Campylobacter* where as high as 20 % in both poultry isolates and those isolates cultured from people.

2.7 When little or No Fluoroquinolone is Used in Food Animals There is Little Fluoroquinolone Resistance

In countries that have never allowed the use of fluoroquinolones in food animals e.g. Australia there is almost no resistance seen in *E. coli*, *Salmonella*, or *Campylobacter* in isolates derive from food animals or in foods produced from these animals

(Collignon et al. 2008a, b; Unicomb et al. 2003). It thus appears very likely that the major driving factor for resistance in most of these food borne pathogens that are derived from animals, is the use of antimicrobials and the types of the antimicrobials use in food animals.

In children and in pregnant women, fluoroquinolones are contraindicated and thus for invasive or serious disease with *Salmonella*, third generation cephalosporins are the agents of choice. Unfortunately increasing rates of resistance in *Salmonella* make this option difficult. This is particularly a problem in developing countries where invasive *Salmonella* infections are much more common. It, however, is also a problem for those living in developed countries as these infections can be acquired domestically and also by travellers when they have visited countries with much higher endemic rates of infections and/or resistance. In many countries (e.g. Denmark) imported foods have on them bacteria that are much more resistant than found on domestically produced foods (DANMAP 2009).

2.8 What Do We Need To Do?

We need to better control the development of resistant bacteria in people by considerably lowering the volumes of antibiotics we use. In most countries we need to reduce by at least 50 % or more, the total amount of antibiotics used in people as the majority is used for viral infections or is ineffectual. We also need to limit the spread of resistant bacteria by better hygiene and infection control. Otherwise resistance inevitably rises and rises.

However, even if we had both these optimally processes performed in people, this would not solve the problem. Resistant bacteria develop wherever antibiotics are used and two-thirds or more of all the antibiotics used in the world are used in food production animals. Aquaculture is also rapidly expanding and so it their use of antibiotics. Thus when resistant bacteria develop in these sectors they inevitably also spread to people. It profoundly complicates the issue of optimally managing antibiotic resistance if all the focus is just from a medical perspective in the human health sector.

This is where the concept of One Health is so important. We recognise that one sector impinges on the health of other sectors. It then follows that if we cannot only lower the usage of antibiotics in people but also better control and significantly lower the amounts of antibiotics used in food animals and aquaculture, this will then have a major flow-on effect to the human sector. We also need to also look at the water and waste from these animals and people, as this water will inevitably be contaminated with resistant bacteria and this water will be ingested by people and animals.

Interventions, particularly those targeting better infection control and improved antimicrobial use (decreasing use of broad spectrum antibiotics especially cephalosporins and fluoroquinolones) have made a difference. Antimicrobial stewardship (involving education plus restrictions on the types and quantities of antimicrobial used) is a major way that antibiotic usage can be improved. Attempts to implement this successfully are occurring. However, both implementation and sustainability are difficult. It can also involve significant increased expenditure especially if electronic prescribing and data collection are part of the improved process.

The major action items needed are:

- Control and limit the amounts and types of antibiotic used in people.
- Control and limits the amounts and types of antibiotic used in non-human sectors. This is particularly in food animals and aquaculture but also in areas such as horticulture where heat stable compounds such as gentamicin and streptomycin can be used to spray apples.
- Better infection control and hygiene in people so that even if resistant bacteria develop, we better limit the spread to other people.
- Prevent infections by effective and safe vaccines.
- Clean water. Water in many countries, particularly in the developing world is a major vehicle that allows the spread of resistant bacteria from person to person, animal to animal and animal to people. We need to keep contaminated animal waste and human waste out of water ways as much as is practicable and ensuring water is treated to an appropriate standard.
- No multiantibiotic resistant bacteria in our foods. This will be best achieved by stopping the use of "critically important" antibiotics in our food animals and better limiting the use of all antibiotics in food animals. Globalisation of food can then spread these bacteria widely.
- Better controls on how animal and human faecal waste manure is used to help grow food. This waste will contain large numbers of pathogens including resistant bacteria that may not be inactivated or removed before the foods reach the market.
- Better surveillance of antimicrobial usage and of resistant bacteria—locally, nationally and internationally.

3 Conclusions

Antibiotic resistance causes increased deaths, complications, expenses, prolonged hospital stays, toxicity and difficulties in delivering therapy in the safest way to patients. Antibiotic resistance is a continuing and growing problem. There are not many likely new classes of antibiotics that will be available in the next few decades. Thus we need to preserve the usefulness of those antibiotics we currently have by decreasing their overall use, and especially the use of broad spectrum agents. We also need to improve our ability to prevent infections and the spread of resistant bacteria wherever they arise or are found. This means improving our practices with infection control, hygiene, animal husbandry and development and delivery of effective and safe vaccines. We need clean water to be available for people and for animals. Failure to do this will result in huge numbers of people entering a "post-antibiotic era" for too many common infections.

References

Aarestrup FM, Wegener HC, Collignon P (2008a) Resistance in bacteria of the food chain: epidemiology and control strategies. Expert Rev Anti Infect Ther 6:733–750

Aarestrup FM, McDermott PF, Wegener HC (2008b) Transmission of antimicrobial resistance from food animals to humans. In: Nachamkin I, 25, Szymanski C, Blaser M, (eds) *Campylobacter* 3rd edn. ASM Press, Washington DC, pp 645–665

Beidenbach DJ, Moet GJ, Jones RN (2004) Occurrence and antimicrobial resistance patterns comparisons among bloodstream infection isolates from the SENTRY antimicrobial surveillance programme (1997–2002). Diagn Microbiol Infect Dis 50:59–69

Brinas L, Moreno MA, Zarazaga M, Porrero C, Sáenz Y, García M, Dominguez L, Torres C (2003) Detection of CMY-2, CTX-M-14, and SHV-12 beta-lactamases in *Escherichia coli* fecal-sample isolates from healthy chickens. Antimicrob Ag Chemother 47:2056–2058

Canadian Integrated Program for Antimicrobial Resistance Surveillance (CIPARS) (2007) *Salmonella* Heidelberg ceftiofur-related resistance in human and retail chicken isolates. Modified 16 Aug 2007. Available at www.phac-aspc.gc.ca/cipars-picra/heidelberg/heidelberg-eng.php

Carlet J, Collignon P, Goldmann D, Goossens H, Harbarth S, Jarlier V, Levy SB, N'Doye B, Pittet D, Richtmann R, Seto WH, van der Meer JW, Voss A (2011) Society's failure to protect a precious resource: antibiotics. Lancet 378:369–371

Cavaco LM, Abatih E, Aarestrup FM, Guardabassi L (2008) Selection and persistence of CTX-M-producing *Escherichia coli* in the intestinal flora of pigs treated with amoxicillin, ceftiofur or cefquinome. Antimicrob Ag Chemother 52:3612–3616

CDC Centers for Disease Control and Prevention (2011) Investigation update: outbreak of Shiga toxin-producing *E. coli* O104 (STEC O104:H4) infections associated with travel to Germany. Updated 8 July 2011 (FINAL Update), CDC, Atlanta. Department of Health and human services. Available at http://www.cdc.gov/ecoli/2011/ecolio104/

Collignon P, Courvalin P, Aidara-Kane A (2008a) Clinical importance of antimicrobial drugs to human health. Chapter 4. In: Guardabassi L, Jensen L, Kruse H (eds). Guide to antimicrobial use in animals, Blackwell, Oxford

Collignon P, Dreimanis D, Ferguson J (2008b) Bloodstream infections. In: Cruickshank M, Ferguson J (eds), Reducing harm to patients from health care associated infection: the role of surveillance: Australian Commission on Safety and Quality in Health Care, pp 53–89. Available at www.health.gov.au/internet/safety/publishing.nsf/Content/prog-HAI_Surveillance

Collignon P, Nimmo GR, Gottlieb T, Gosbell IB (2005) Australian Group on antimicrobial resistance. *Staphylococcus aureus* bacteremia, Australia. Emerg Infect Dis 11:554–561. Available at www.cdc.gov/ncidod/eid/vol11no04/pdfs/04-0772.pdf

Collignon PJ, Turnidge JD (2000) Antibiotic resistance in *Streptococcus pneumoniae*. Med J Aust 173(Suppl):S58–S64

Collignon P, Powers JH, Chiller TM, Aidara-Kane A, Aarestrup FM (2009) World Health Organization ranking of antimicrobials according to their importance in human medicine: a critical step for developing risk management strategies for the use of antimicrobials in food production animals. Clin Infect Dis 49(1):132–141, 1 July

Collignon P, Angulo FJ (2006) Fluoroquinolone-resistant *Escherichia coli*: food for thought. J Infect Dis 194(1):8–10

Decousser JW, Pina P, Picot F et al. Delalande C, Pangon B, Courvalin P, Allouch P, ColBVH Study Group (2003). Frequency of isolation and antimicrobial susceptibility of bacterial pathogens isolated from patients with bloodstream infections: a French prospective national survey. J Antimicrob Chemother 51:1213–1222

Corpet DE (1988) Antibiotic resistance from food. New Engl J Med 318:1206–1207

Cosgrove SE, Sakoulas G, Perencevich EN, Schwaber MJ, Karchmer AW, Carmeli Y (2003) Comparison of mortality associated with methicillin-resistant and methicillin-susceptible *Staphylococcus aureus* bacteremia: a meta-analysis. Clinical Infect Dis 36:53–59

Danish Integrated Antimicrobial resistance Monitoring and Research Programme (DANMAP) (2009) Use of antimicrobial agents and occurrence of antimicrobial resistance in bacteria from food animals, foods and humans in Denmark. Statens Serum Institut, Danish Veterinary and Food Administration, Danish Medicines Agency and Danish Institute for Food and Veterinary Research, Copenhagen, 2010. Available at www.danmap.org/pdfFiles/Danmap_2009.pdf

Davies J, Davies D (2010) Origins and evolution of antibiotic resistance. Microbiol Mol Biol Rev 74(3):417–433

de Kraker ME, Wolkewitz M, Davey PG, Koller W, Berger J, Nagler J, Icket C, Kalenic S, Horvatic J, Seifert H, Kaasch A, Paniara O, Argyropoulou A, Bompola M, Smyth E, Skally M, Raglio A, Dumpis U, Melbarde Kelmere A, Borg M, Xuereb D, Ghita MC, Noble M, Kolman J, Grabljevec S, Turner D, Lansbury L, Grundmann H (2011) Burden of antimicrobial resistance in European hospitals: excess mortality and length of hospital stay associated with bloodstream infections due to *Escherichia coli* resistant to third-generation cephalosporins. J Antimicrob Chemother 66:398–407

Diwan V, Tamhankar AJ, Khandal RK, Sen S, Aggarwal M, Marothi Y, Iyer RV, Sundblad-Tonderski K, Stålsby-Lundborg C (2010) Antibiotics and antibiotic-resistant bacteria in waters associated with a hospital in Ujjain, India. BMC Public Health 13(10):414

Engberg J, Aarestrup FM, Taylor DE, Gerner-Smidt P, Nachamkin I (2001) Quinolone and macrolide resistance in *Campylobacter jejuni* and *C. coli*: resistance mechanisms and trends in human isolates. Emerg Infect Dis 7:24–34

European Centre for Disease Prevention and Control (ECDC) (2010). Antimicrobial resistance surveillance in Europe 2009. Annual Report of the European Antimicrobial Resistance Surveillance Network (EARS-Net). ECDC, Stockholm. Available at www.ecdc.europa.eu/en/publications/Publications/1011_SUR_annual_EARS_Net_2009.pdf

Food and Agriculture Organization of the United Nations (FAO)/World Organisation for Animal Health (OIE)/World Health Organization (WHO) (2003) Joint expert workshop on non-human antimicrobial usage and antimicrobial resistance: scientific assessment. WHO, Geneva. Available at www.who.int/foodsafety/publications/micro/en/amr.pdf Accessed 1–5 Dec

Finland M, Jones WF Jr, Barnes MW (1959) Occurrence of serious bacterial infections since introduction of antibacterial agents. J Am Med Assoc 170:2188–2197

Fernando GA, Collignon PJ, Bell JM (2010) A risk for returned travellers: the "post-antibiotic era". Med J Aust. 10(193):59

Health Protection Agency (2009) Quarterly results from the mandatory surveillance of MRSA bacteraemia. MRSA bacteraemia, Health Protection agency, June. Available at www.hpa.org.uk/web/HPAweb&HPAwebStandard/HPAweb_C/1233906819629

Heuer OE, Hammerum AM, Collignon P, Wegener HC (2006) Human health hazard from antimicrobial-resistant enterococci in animals and food. Clin Infect Dis 43:911–916

Ho PL, Chow KH, Lai EL, Lo WU, Yeung MK, Chan J, Chan PY, Yuen KY (2011) Extensive dissemination of CTX-M-producing *Escherichia coli* with multidrug resistance to 'critically important' antibiotics among food animals in Hong Kong, 2008–2010. J Antimicrob Chemother 66(4):765–768

Hsueh P-R, Teng L-J, Lee L-N, Yang PC, Ho SW, Luh KT (1999) Dissemination of high-level penicillin-, extended-spectrum cephalosporin-, and erythromycin-resistant *Streptococcus pneumoniae* clones in Taiwan. J Clin Microbiol 37:221–224

Huijsdens XW, van Dijke BJ, Spalburg E, van Santen-Verheuvel MG, Heck ME, Pluister GN, Voss A, Wannet WJ, de Neeling AJ (2006) Community-acquired MRSA and pig-farming. Ann Clin Microbiol Antimicrob 5:26

Hujer KM, Hujer AM, Hulten EA, Bajaksouzian S, Adams JM, Donskey CJ, Ecker DJ, Massire C, Eshoo MW, Sampath R, Thomson JM, Rather PN, Craft DW, Fishbain JT, Ewell AJ, Jacobs MR, Paterson DL, Bonomo RA (2006) Analysis of antibiotic resistance genes in multidrug-resistant Acinetobacter sp. isolates from military and civilian patients treated at the Walter Reed Army Medical Center. Antimicrob Ag Chemother 50(12):4114–4123

Iovine NM, Blaser MJ (2004) Antibiotics in animal feed and spread of resistant *Campylobacter* from poultry to humans. Emerg Infect Dis 10:1158–1159

JETACAR (1999) The use of antibiotics in food-producing animals. Antibiotic-resistant bacteria in animals and humans 1999. Available at www.health.gov.au/internet/main/Publishing.nsf/Content/2A8435C711929352CA256F180057901E/$File/jetacar.pdf

Johnson JR, Kuskowski MA, Menard M, Gajewski A, Xercavins M, Garau J (2006) Similarity of human and chicken *Escherichia coli* isolates with relation to ciprofloxacin resistance status. J Infect Dis 2006(194):71–78

Kennedy KJ, Roberts JL, Collignon PJ (2008) *Escherichia coli* bacteraemia in Canberra: incidence and clinical features. Med J Aust 188:209–213

Kennedy K, Collignon P (2010) Colonisation with *Escherichia coli* resistant to "critically important" antibiotics: a high risk for international travellers. Eur J Clin Microbiol Infect Dis 29(12):1501–1506

Khanna T, Friendship R, Dewey C, Weese JS (2008) Methicillin resistant *Staphylococcus aureus* colonization in pigs and pig farmers. Vet Microbiol 128:298–303

Klevens RM, Morrison MA, Nadle J, Petit S, Gershman K, Ray S, Harrison LH, Lynfield R, Dumyati G, Townes JM, Craig AS, Zell ER, Fosheim GE, McDougal LK, Carey RB, Fridkin SK (2007) Active bacterial core surveillance (ABCs) MRSA investigators. Invasive methicillin-resistant *Staphylococcus aureus* infections in the United States. J Am Med Assoc 298:1763–1771

Kumarasamy KK, Toleman MA, Walsh TR, Bagaria J, Butt F, Balakrishnan R, Chaudhary U, Doumith M, Giske CG, Irfan S, Krishnan P, Kumar AV, Maharjan S, Mushtaq S, Noorie T, Paterson DL, Pearson A, Perry C, Pike R, Rao B, Ray U, Sarma JB, Sharma M, Sheridan E, Thirunarayan MA, Turton J, Upadhyay S, Warner M, Welfare W, Livermore DM, Woodford N (2010) Emergence of a new antibiotic resistance mechanism in India, Pakistan, and the UK: a molecular, biological, and epidemiological study. Lancet Infect Dis 2010(9):597–602

Lewis HC, Mølbak K, Reese C, Aarestrup FM, Selchau M, Sørum M, Skov RL (2008) Pigs as Source of methicillin-resistant *Staphylococcus aureus* CC398 infections in humans. Den Emerg Infect Dis 14:1383–1388

Li J, Nation RL, Turnidge JD, Milne RW, Coulthard K, Rayner CR, Paterson DL (2006) Colistin: the re-emerging antibiotic for multidrug-resistant gram-negative bacterial infections. Lancet Infect Dis 6:589–601

Mayerhofer G, Schwaiger-Nemirova I, Kuhn T, Girsch L, Allerberger F (2009) Detecting streptomycin in apples from orchards treated for fire blight. J Antimicrob Chemother 63(5):1076–1077

Markestad A, Grave K (1997) Reduction of antibacterial drug use in Norwegian fish farming due to vaccination. Dev Biol Stand 90:365–369

Mead PS, Slutsker L, Dietz V, McCaig LF, Bresee JS, Shapiro C, Griffin PM, Tauxe RV (1999) Food-related illness and death in the United States. Emerg Infect Dis 5:607–625

Mesa RJ, Blanc V, Blanch AR, Cortés P, González JJ, Lavilla S, Miró E, Muniesa M, Saco M, Tórtola MT, Mirelis B, Coll P, Llagostera M, Prats G, Navarro F (2006) Extended-spectrum beta-lactamase-producing Enterobacteriaceae in different environments (humans, food, animal farms and sewage). J Antimicrob Chemother 58:211–215

Moellering R (2005) *Enterococcus* species, *Streptococcus bovis* and *Leuconostoc* species. In: Mandell, GL, Bennett JE, Dolin R (eds) Principles and practice of infectious diseases 6th edn. Elsevier Churchill Livingstone, Philadelphia pp 2411–2422

Pallares R, Liñares J, Vadillo M, Cabellos C, Manresa F, Viladrich PF, Martin R, Gudiol F (1995) Resistance to penicillin and cephalosporin and mortality from severe pneumococcal pneumonia in Barcelona, Spain. New Engl J Med 333:474–480

Pegues D, Ohl M, Miller S (2005) *Salmonella* species including *Salmonella typhi*. In: Mandell, GL, Bennett JE, Dolin R (eds) Principles and practice of infectious diseases 6th edn. Elsevier Churchill Livingstone, Philadelphia pp 2636–2654

Power E (2006) Impact of antibiotic restrictions: the pharmaceutical perspective. Clin Microbiol Infect 12(Suppl 5):25–34

Tängdén T, Cars O, Melhus A, Löwdin E (2010) Foreign travel is a major risk factor for colonization with *Escherichia coli* producing CTX-M-type extended-spectrum beta-lactamases: a prospective study with Swedish volunteers. Antimicrob Ag Chemother 54:3564–3568

Tumbarello M, Spanu T, Di Bidino R, Marchetti M, Ruggeri M, Trecarichi EM, De Pascale G, Proli EM, Cauda R, Cicchetti A, Fadda G (2010) Costs of bloodstream infections caused by

Escherichia coli and influence of extended-spectrum-beta-lactamase production and inadequate initial antibiotic therapy. Antimicrob Ag Chemother 54:4085–4091

Unicomb L, Ferguson j, Riley TV, Collignon P (2003) Fluoroquinolone resistance in *Campylobacter* absent from isolates, Australia. Emerg Infect Dis 2003(9):1482–1483

Walsh TR, Weeks J, Livermore DM, Toleman MA (2011) Dissemination of NDM-1 positive bacteria in the New Delhi environment and its implications for human health: an environmental point prevalence study. Lancet Infect Dis 11:355–362

Wang H, Chen M (2005) China Nosocomial Pathogens Resistance Surveillance Study Group. Surveillance for antimicrobial resistance among clinical isolates of gram-negative bacteria from intensive care unit patients in China, 1996 to 2002. Diagn Microbiol Infect Dis 51:201–208

Webb V, Davies J (1993) Antibiotic preparations contain DNA: a source of drug resistance genes? Antimicobiol Ag Chemother 37:2379–2384

World Health Organization (WHO) (2009). WHO list of Critically Important Antimicrobials (CIA). 2nd Rev. WHO Advisory Group on Integrated Surveillance of Antimicrobial Resistance (AGISAR), Copenhagen, 2009. Available at www.who.int/foodborne_disease/resistance/cia/en/index.html

Zhao S, White DG, McDermott PF, Friedman S, English L, Ayers S, Meng J, Maurer JJ, Holland R, Walker RD (2001) Identification and expression of cephamycinase bla(CMY) genes in *Escherichia coli* and *Salmonella* isolates from food animals and ground meat. Antimicrob Ag Chemother 45:3647–3650

Zhou LJ, Ying GG, Zhao JL, Yang JF, Wang L, Yang B, Liu S (2011) Trends in the occurrence of human and veterinary antibiotics in the sediments of the Yellow River, Hai River and Liao River in northern China. Environ Pollut 159(7):1877–1885

Bovine Spongiform Encephalopathy: A Tipping Point in One Health and Food Safety

James Hope

Abstract Bovine spongiform encephalopathy (BSE) is a protein misfolding disease of cattle which belongs to the group of transmissible spongiform encephalopathies (TSEs) or prion diseases. This group also includes scrapie in sheep and goats, chronic wasting disease (CWD) of cervids and Creutzfeldt-Jakob disease (CJD) humans. The first case of BSE was recognised in England in 1986 as a progressive, neurological condition where affected animals behaved abnormally, exhibited anxiety, ataxia, hypersensitivity to touch and noise and poor body condition. Spongiform change was observed in the brain stem of cattle at post mortem and its similarity to scrapie in sheep stimulated biochemical investigation and transmission studies which confirmed it as a novel prion disease of cattle. Epidemiological analysis of the initial cases of disease implicated a common extended source of infection, likely to be related to feed, and stimulated a series of control measures designed to restrict feeding of mammalian-derived protein to ruminants in various parts of the United Kingdom and to prevent the use of various bovine offals in feed or food production. This article outlines the rise and fall of the incidence of BSE in the UK and Europe, its classification as a zoonotic disease with the emergence of variant CJD, the implications of it as a prion disease and challenge its diagnosis and control continues to represent worldwide.

Contents

J. Hope (✉)
Animal Health and Veterinary Laboratories Agency Weybridge,
Woodham Lane, New Haw, Addlestone, KT15 3NB, Surrey, UK
e-mail: James.Hope@ahvla.gsi.gov.uk

Current Topics in Microbiology and Immunology (2013) 366: 37–47
DOI: 10.1007/82_2012_264

Published Online: 14 September 2012

1 Prions: The BSE Agent

Prions are proteins encoded by genes in chromosomal DNA that when expressed in cells can transmit information in and between cells and their parent organisms by virtue of variable folding of their polypeptide chain. Frequently, the conversion of a normal cellular protein into a prion form involves a conformational change that affects its degree of self-association and ability to interact with other molecules (Wickner et al. 2009). In yeast and fungi, prion forms of particular proteins are the molecular basis of a type of extra-chromosomal inheritance of survival traits but in mammals they were first investigated as the causative agents, as well as by-products, of a family of progressive, neurological degenerative disease known as the transmissible spongiform encephalopathies (TSEs). TSEs are diseases characterised by the accumulation of the prion form of the mammalian prion protein (PrP) in the central nervous system or peripheral tissues of animals and humans (Prusiner 1997). Scrapie is the TSE of sheep and goats; Creutzfeldt-Jakob disease (CJD) is the most common type of human TSE, and a novel variant (vCJD) is believed to be caused by the transmission of the cattle TSE, bovine spongiform encephalopathy (BSE) to humans.

2 A Brief History of BSE

Bovine spongiform encephalopathy (BSE) has adversely affected the UK cattle industry, and inhibited trade in bovine-derived products and animal feed worldwide, for the past 25 years. The first case of BSE was recognised retrospectively to have been seen in April 1985, but it was a further 20 months, in December 1986, before cases of this spongiform encephalopathy in cattle were confirmed by pathologists at the Central Veterinary Laboratory, Weybridge (Wells et al. 1987). By that time, over 100,000 cattle were infected with this long-incubation period disease which cannot be detected by serology, polymerase chain reaction or by other ante-mortem testing and a major epidemic of clinical cases developed which peaked in the British Isles in 1993 with over 37,000 cases diagnosed that year ((www.defra.gov.uk/vla/science/sci_tse_stats); Fig. 1). Dairy cows appeared at higher risk than beef cattle but either sex could be affected usually within

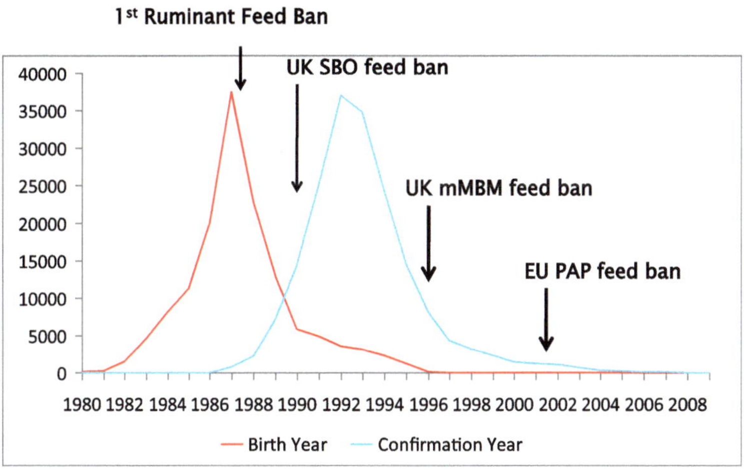

Fig. 1 BSE epidemic curves and the effect of feed bans for the United Kingdom

4–4.5 years of birth (range 1.8–18 years). Most early cases of BSE occurred in cattle between the ages of 3 and 5 years in Great Britain, but as the epidemic has waned and level of exposure declined, the average age of cattle with confirmed signs of disease has increased gradually to over 13 years (EFSA 2009). The year of peak incidence in Europe has varied from country to country and reflects the staggered nature of the BSE epidemics that have spread throughout Europe from the UK. For most of its development time the disease produces not overt clinical signs (Wilesmith et al. 1988) and the inability to detect the asymptomatic carrier of BSE limits refinement of the measures which can be taken to prevent infected bovine tissues from use in feed and pharmaceutical products.

3 Feed Bans and Born-After-Reinforced-Ban Cases

Epidemiological analyses of BSE-affected herds identified a protein feed supplement to be the most likely source of infection (Wilesmith et al. 1988), and subsequent recycling of BSE-infected cattle waste in this process may have contributed to the persistence of the disease. Ruminant feed legislation aimed at removing the source of infection from cattle born after 1988 was introduced in 1989–1990 in the UK, with more stringent measures in 1996, and these bans were re-inforced throughout Europe in 2001. The initial UK ban was aimed at prohibiting "cannabalism" by ruminants, end the practice of feeding ruminant-derived meat and bone meal (MBM) to ruminants and so preventing the recycling and amplification of prions in cattle via rendered material. This worked (see Fig. 1) but, geographically, its effect was more noticeable in the south and west than in the north and east of England, the latter area with a higher proportion of pigs to cattle and sheep. This lead to the suspicion that pig and poultry rations (in which ruminant MBM was allowed) were finding their way into ruminant feed and in November 1994 it was prohibited to feed any mammalian protein (with a few exceptions—e.g. blood and milk proteins) to ruminants. After the reporting of a link between BSE and vCJD in 1996, the ban was immediately extended to make illegal the feeding of any mammalian MBM to any farmed livestock, including fish and horses. Ever more stringent feed controls were imposed throughout Europe to limit the spread of BSE within the European Union trade zone and a complete ban on feeding processed animal protein (PAP) of any origin to animals kept, fattened or bred for the production of food was introduced in 2001. Not all countries at risk of BSE worldwide introduced similar exhaustive bans and part of the current dilemma faced by the EU in trying to relax its legislation, in proportion to the diminishing numbers of BSE cases in Europe, is the fear of re-introduction of BSE prions by import of infected material from outside its borders. Feeding pig and poultry processed animal protein to fish is currently under consideration and there is a continuing (unresolved) debate within member states on the use of poultry PAP in pig feed and pig PAP in poultry feed although the risk of the re-amplification of BSE by this usage appears to be vanishingly small (EFSA 2011). Cannabalism, that is feeding PAP of one species to the same species, is widely resisted and there appears to be no European agenda for the re-introduction of feeding ruminant PAP to any species.

At its peak, over 1,000 clinical cases were reported each week in Great Britain in 1993 and the last clinical case was seen by passive surveillance in 2009 (Fig. 1). Active surveillance for disease in fallen stock, in healthy slaughter animals (born after the 1996 feed ban) at the abattoir and in various cohort culls was introduced in 2001 in the UK (and in other member states of the European Union) and, in 2011 when some 500,000 animals were tested, the number of cases detected (all in fallen stock) had dwindled to 7. To date (June 2012), in the UK, only one animal has tested positive for BSE in 2012. Although the feed bans have had a dramatic effect on the epidemic curve, cases of BSE continued to be confirmed in cattle born after the re-inforced European bans of 2000 and feeding of contaminated protein to calves continues to be suspected as the reason for most of these 'born after the real ban' (BARB) cases (Ortiz-Pelaez et al. 2012). The biological and biochemical characteristics of BARBs appear similar to those seen in earlier cases in the epidemic, although two of the most recent BARB cases in Britain, including the 2012 case, have had the molecular properties of L-type atypical BSE (see below, Stack et al. in press).

In parallel with the BSE epidemic, natural cases of transmissible spongiform encephalopathies have been also reported for the first time in cattle-related species—greater kudu, eland, nyala and gemsbok, Arabian and scimitar-horned oryx (Cunningham et al. 2004) and in the cat family -puma, cheetahs- and domestic cats. Apart from some cases in the greater kudu, contaminated feed is suspected but difficult to prove because of the absence of detailed feeding records (Cunningham et al. 2004).

4 Minimum Effective Dose of BSE Prions and Specified Risk Material Controls

Experimental oral dosing of cattle with BSE-affected cattle brain homogenates has confirmed that as little as 1 milligram of brain (with ~10–100 mouse ic ID_{50} units) can induce disease after extended incubation periods of 8–10 years (Arnold et al. 2007, 2009; Wells et al. 2007). Larger doses (up to 100 g) have been used to study oral pathogenesis of BSE in cattle in the UK (Wells et al. 2007) and Germany (Hoffmann et al. 2007). These studies have confirmed by PrP IHC and bioassay the limited, early distribution of prions to parts of the lower alimentary tract (distil ileum, jejunum) and spread via the autonomic nervous system from the gastrointestinal tract to the central nervous system via either the coeliac and mesenteric ganglion complex, splanchnic nerves and the lumbal/caudal thoracic spinal cord or via the vagal nerve. This experimental tissue distribution of infectivity has been used to refine the list of specified risk materials from various age-cohorts of cattle banned for human consumption and has underpinned several assessments of human and animal exposure risk that have defined UK and European policy for control and management of BSE over the years, such as a recent EFSA Opinion on the BSE risk of bovine intestines (EFSA 2010).

5 Atypical Forms of BSE

BSE surveillance testing of cattle for abnormal prion protein in Europe has allowed the identification of two further, distinct types of cattle TSE, termed H- and L-(or BASE) type BSE (Casalone et al. 2004; Biacabe et al. 2007; Jacobs et al. 2007; Polak et al. 2008). Similar cases were also detected outside Europe (Japan and USA) (Hagiwara et al. 2007; Clawson et al. 2008). About 60 atypical BSE cases have been described worldwide (from testing ~50 million healthy animals and fallen stock) although there is no statutory requirement to distinguish typical and atypical types of BSE in reporting and this figure is derived from research literature.

In France, a retrospective study of all the TSE-positive cattle identified through the compulsory EU surveillance programme between 2001 and 2007 was recently published (Biacabe et al. 2008). This study indicated that all BSE H and L cases detected by rapid tests were observed in animals over 8 years old in either the "at risk" (9) or "healthy slaughtered" surveillance target group (4). In this study, the reported frequency of H and L type TSE was respectively 1.9 and 1.7 cases per million of over 8-year-old tested animals. All EU atypical cases were born before the extended or real feed ban that came into law in January 2001. Hence, as with classical BSE, exposure of these animals to feed contaminated with low titres of TSE cannot be excluded. However, the distribution of H-and L-type cases in France by year of birth differs markedly from that for classical BSE and could be interpreted to indicate that both forms of atypical BSE are sporadic diseases which arise spontaneously. Indeed, a case of H-type BSE in the USA has been associated with a inheritable, bovine E211 K mutation (that is the amino-acid lysine replacing glutamic acid at codon 211) in the wild-type prion protein amino-acid sequence and shown to be transmissible by intra-cranial inoculation to a calf carrying the same mutated allele with a post-inoculation survival time of 301 days (Greenlee et al. 2012); however, DNA sequencing of the PrP open reading frame of other atypical cases has not identified this or any other coding region mutation.

H- and L-(or BASE) type BSE have been transmitted by intra-cerebral challenge to inbred mice and Tg mice expressing bovine and ovine PrP. L-type BSE has also been transmitted to transgenic mice expressing alleles of the human prion protein (Buschmann et al. 2006; Beringue et al. 2008; Kong et al. 2008). Transmission and serial passage in inbred mice and Tg VRQ mice have been interpreted to indicate that, after interspecies passage, BASE could generate classical BSE (Beringue et al. 2007; Capobianco et al. 2007). However, it should be noted that L-BSE—classical BSE phenotypic convergence has not been observed in other Tg mice, including mice expressing the ARQ allele of sheep PrP (Buschmann et al. 2006; Beringue et al. 2007). Prions with the properties of a classical BSE strain also emerged during serial passage of H-type BSE in wild-type mice (Baron et al. 2011). These phenomena need to be confirmed in an independent set of experiments but do raise the issue of a possible classical BSE re-emergence originating from atypical BSE cases; mutation rates of 1 in 100,000 have been calculated for conversion (or re-version) of one prion strain to another during passage in cell culture (Oelschlegel et al. 2012; Weissmann 2012) but a similar estimate of the probability of one type of prion changing to another (for example, from a non-zoonotic scrapie to zoonotic BSE phenotype) has not been determined for either experimental or natural transmission routes in animals.

The sensitivity and specificity of the TSE rapid screening tests are known for classical BSE but not for H- or L-type BSE. These tests use brainstem as the target tissue because this is where pathological lesions and PrPres are first detected in the CNS of cattle (Hope et al. 1988; Wells et al. 1998). Unlike classical BSE, little is known about the pathogenesis of atypical BSE and the brainstem may not be the optimal target site for the detection of H- and L- type BSE (Casalone et al. 2004). Consequently the BSE H- and L- type prevalence of 1–2 per million may be an under-estimation. H-type and L-type BSE have been transmitted to cattle (Lombardi et al. 2008; Fukuda et al. 2009) and the molecular and pathological characteristics of each type has been maintained and differ from those of classical BSE (Balkema-Buschmann et al. 2011; Okada et al. 2011a, b; Konold et al. 2012). Some data are now emerging on the distribution of the infectivity in peripheral tissues of cattle with atypical BSE (Iwamaru et al. 2010; Okada et al. 2011a, b; Suardi et al. 2012) and, although limited compared to the distribution of prions in small ruminants, various skeletal muscles in L-type (BASE) cattle were found to contain infectivity (detected by bioassay in Tgbov mice) and PrP-immuno-reactive deposits within individual muscle fibres (Suardi et al. 2012).

6 Small Ruminant BSE

Foster and colleagues showed cattle BSE could be transmitted to ARQ/ARQ sheep and goats by feeding and intra-cerebral inoculation (Foster et al. 1993) and several subsequent studies have documented that there is wide-spread dissemination of BSE prions in ARQ/ARQ sheep similar to the pathogenesis of natural cases of classical scrapie (van Keulen et al 2000). The biological and biochemical characteristics of "BSE in small ruminants" are sufficiently distinct to allow their discrimination in "blinded" tests although there have been concerns "mixed" infections might pass as "scrapie". Historically, small ruminants were known to have been fed the same type of protein supplements implicated as the source of BSE in cattle and fear of a second wave of vCJD due to infection from sheep and goat products stimulated intensive surveillance in the EU of TSEs in sheep and goats and the application of laboratory tests aimed at a diagnosis of "NOT BSE" or "BSE NOT Excluded"; the final confirmation of "BSE in small ruminant" requires the application of bioassay in the same panels of inbred mice used to characterise vCJD and BSE (Bruce et al. 1997). By these stringent criteria, only two cases of BSE in small ruminants(SRs), both in goats, have been confirmed (Eloit et al. 2005; Jeffrey et al. 2006) and current estimates of the likely prevalence of BSE in SRs in Europe is very low.

7 Variant Creutzfeldt-Jakob Disease

In April 1996, Will and colleagues (Will et al. 1996) reported a novel variant of CJD (vCJD). The initial, and subsequent, focus of vCJD in Great Britain and its molecular (Collinge et al. 1996) and transmission (Lasmezas et al. 1996; Bruce et al. 1997) similarities to BSE immediately implicated the cattle disease as the source of vCJD infection (Fig. 2) and beef and cattle by-products were put under further restrictions to limit the spread of disease. Nevertheless, an estimated three million infected cattle may have entered the human food chain (Ghani et al. 2000) and the impact and cost of preventing a secondary, human-to-human wave of infection are still being felt in the UK (Garske and Ghani 2010).

To date (June 2012), there have been 176 primary cases, and three secondary cases related to transfusion of blood products, in the UK, 26 cases in France, 5 in Spain and 16 in the rest of Europe; other cases have been reported in the United States, Canada, Saudi Arabia, Taiwan and Japan (www.cjd.ed.ac.uk/vcjdworld). Polymorphisms and mutations in the human prion protein gene are known to affect the survival and clinico-pathological phenotype of human TSEs and the vCJD is clearly associated with the dimorphism at codon 129 (Methionine or Valine): the percentage of this dimorphism in the normal Caucasian population is 39 MM, 50 MV and 11 % VV; in cases of sporadic CJD, the proportions are 65 MM, 17 MV and 18 % VV. All clinical cases of vCJD so far diagnosed are MM homozygous

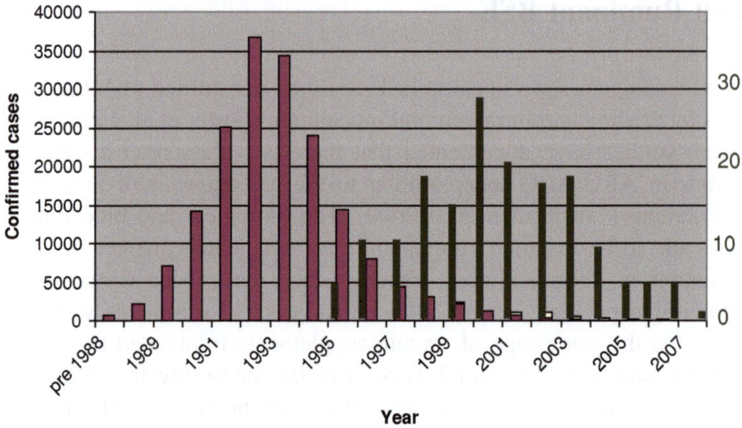

Fig. 2 BSE and vCJD in the UK: temporal link. Left-hand scale, thick bars, BSE cattle; right-hand scale, thin bars, vCJD cases

although pre/sub-clinical disease has been inferred from PrP^{CJD} detection in the spleen of a blood transfusion recipient who was an M/V heterozygote (Peden et al. 2010).

The true extent of vCJD in the UK population can only be guessed at from the low incidence of clinical disease and several large-scale surveys of lymphoid tissue—tonsils or appendices—for biochemical evidence of abnormal prion protein (PrP^{CJD}) by IHC or ELISA have been carried out to provide some support for estimates of disease prevalence. Investigation of the spread and replication of prions in animals has implicated the central role of lympho-reticular tissue in the pathogenesis of these diseases, and abnormal prion protein has been detected in tonsil and spleen of vCJD patients. In an early study of tonsils and appendices removed by surgery and stored in hospital archives between 1995 and 1999, abnormal prion protein was found in three positive appendix samples out of 12,674 specimens tested suitable for PrP^{CJD} using the immuno-histochemical method. All the PrP^{CJD} positive individuals were in the 20–30-year-old cohort and the prevalence estimate calculated for this age group was 380 per million (95 % CI: 80–1120 per million) (Hilton et al. 2004). This is the "high risk" birth cohort with 80 % of the UK cases of vCJD born between 1961 and 1985.

Each method of analysis has its own sensitivity and specificity and a PrP^{CJD} ELISA failed to detect a single positive specimen in 95,672 tonsils (including ~18,000 from the 1961–1985 "at risk" cohort) from the UK National Anonymous Tonsil Archive (NATA) (Clewley et al. 2009); in a subsequent investigation, PrP^{CJD} IHC detected one single positive in 9,672 tonsils from the 1961 to 1985 birth cohort (de Marco et al. 2010). In the most recent appendix screening study, abnormal prion accumulation was detected within 12–18 appendices out of 32,441 suitable specimens examined, and these IHC positive appendices occurred in all codon-129 genotypes—MM, VV and MV. The estimated overall prevalence of 493 per million (95 % CI: 282–801 per million) was statistically consistent with the earlier results but surprisingly, as only 13 % of vCJD cases were born between 1941 and 1960, the highest prevalence of 733 per million (95 % CI: 269–1596 per million) was observed in this older birth cohort (HPA 2012).

8 One Health

The stringent control measures on feed and SRM removal now in place in Europe have been effective in curbing the epidemic of BSE (and limiting human exposure to prions) but the very nature of the prion as an epigenetic, structural re-arrangement of a protein implies eradication may not be possible and a return to past practice of unregulated feeding of processed animal protein runs the risk of a re-surgence of TSE in food animals. The emergence of BSE and vCJD, and the impact these diseases have made to our perceptions and practice of food security, underline the necessity for a uniform, global approach to this prion dilemma, to sustainability in agriculture and the problem of feeding the population of the world. The One Health concept is ancient but just as relevant today as it has been in the past.

References

Arnold ME, Ryan JB et al (2007) Estimating the temporal relationship between PrPSc detection and incubation period in experimental bovine spongiform encephalopathy of cattle. J Gen Virol 88(Pt 11):3198–3208

Arnold ME, Hawkins SA et al (2009) Pathogenesis of experimental bovine spongiform encephalopathy (BSE): estimation of tissue infectivity according to incubation period. Vet Res 40(1):8

Balkema-Buschmann A, Ziegler U et al (2011) Experimental challenge of cattle with German atypical bovine spongiform encephalopathy (BSE) isolates. J Toxicol Environ Health A 74(2–4):103–109

Baron T, Vulin J et al (2011) Emergence of classical BSE strain properties during serial passages of H-BSE in wild-type mice. PLoS One 6(1):e15839

Beringue V, Andreoletti O et al (2007) A bovine prion acquires an epidemic bovine spongiform encephalopathy strain-like phenotype on interspecies transmission. J Neurosci 27(26):6965–6971

Beringue V, Herzog L et al (2008) Transmission of atypical bovine prions to mice transgenic for human prion protein. Emerg Infect Dis 14(12):1898–1901

Biacabe AG, Jacobs JG et al (2007) H-type bovine spongiform encephalopathy: complex molecular features and similarities with human prion diseases. Prion 1(1):61–68

Biacabe AG, Morignat E et al (2008) Atypical bovine spongiform encephalopathies, France, 2001–2007. Emerg Infect Dis 14(2):298–300

Bruce ME, Will RG et al (1997) Transmissions to mice indicate that 'new variant' CJD is caused by the BSE agent. Nature 389(6650):498–501

Buschmann A, Gretzschel A et al (2006) Atypical BSE in Germany–proof of transmissibility and biochemical characterization. Vet Microbiol 117(2–4):103–116
Capobianco R, Casalone C et al (2007) Conversion of the BASE prion strain into the BSE strain: the origin of BSE? PLoS Pathog 3(3):e31
Casalone C, Zanusso G et al (2004) Identification of a second bovine amyloidotic spongiform encephalopathy: molecular similarities with sporadic Creutzfeldt-Jakob disease. Proc Natl Acad Sci USA 101(9):3065–3070
Clawson ML, Richt JA et al (2008) Association of a bovine prion gene haplotype with atypical BSE. PLoS One 3(3):e1830
Clewley JP, Kelly CM et al (2009) Prevalence of disease related prion protein in anonymous tonsil specimens in Britain: cross sectional opportunistic survey. BMJ 338:b1442
Collinge J, Sidle KC et al (1996) Molecular analysis of prion strain variation and the aetiology of 'new variant' CJD. Nature 383(6602):685–690
Cunningham AA, Kirkwood JK et al (2004) Bovine spongiform encephalopathy infectivity in greater kudu (*Tragelaphus strepsiceros*). Emerg Infect Dis 10(6):1044–1049
de Marco MF, Linehan J et al (2010) Large-scale immunohistochemical examination for lymphoreticular prion protein in tonsil specimens collected in Britain. J Pathol 222(4):380–387
EFSA (2009) Scientific Opinion on the updated risk for human and animal health related to the revision of the BSE monitoring regime in some member states. EFSA Journal 1059:1–40
EFSA (2010) Scientific Opinion on BSE risk in bovine instestines. EFSA Journal 8:1317–1335
EFSA (2011) Scientific opinion on the revision of the quantitative risk assessment (QRA) of the BSE risk posed by processed animal proteins (PAPs). EFSA Journal 9(1):1947–2026
Eloit M, Adjou K et al (2005) BSE agent signatures in a goat. Vet Rec 156(16):523–524
Foster J, Hope J et al (1993) Transmission of bovine spongiform encephalopathy to sheep and goats. Vet Rec 133:339–341
Fukuda S, Iwamaru Y et al (2009) Intraspecies transmission of L-type-like bovine spongiform encephalopathy detected in Japan. Microbiol Immunol 53(12):704–707
Garske T, Ghani AC (2010) Uncertainty in the tail of the variant Creutzfeldt-Jakob disease epidemic in the UK. PLoS One 5(12):e15626
Ghani AC, Donnelly CA et al (2000) Assessment of the prevalence of vCJD through testing tonsils and appendices for abnormal prion protein. Proc Biol Sci 267(1438):23–29
Greenlee JJ, Smith JD et al (2012) Clinical and pathologic features of H-type bovine spongiform encephalopathy associated with E211 K prion protein polymorphism. PLoS ONE 7(6):e38678
Hagiwara K, Yamakawa Y et al (2007) Accumulation of mono-glycosylated form-rich, plaque-forming PrPSc in the second atypical bovine spongiform encephalopathy case in Japan. Jpn J Infect Dis 60(5):305–308
Health Protection Agency Report News Archives (2012) UK Health Protection Agency. Summary results of the second national survey of abnormal prion prevalence in archived appendix specimens 6(29): 20 July 2012. http://www.hpa.org.uk/hpr/archives/2012/news3212.htm#bnrmlp
Hilton DA, Ghani AC et al (2004) Prevalence of lymphoreticular prion protein accumulation in UK tissue samples. J Pathol 203(3):733–739
Hoffmann C, Ziegler U et al (2007) Prions spread via the autonomic nervous system from the gut to the central nervous system in cattle incubating bovine spongiform encephalopathy. J Gen Virol 88(Pt 3):1048–1055
Hope J, Reekie LJ et al (1988) Fibrils from brains of cows with new cattle disease contain scrapie-associated protein. Nature 336(6197):390–392
Iwamaru Y, Imamura M et al (2010) Accumulation of L-type bovine prions in peripheral nerve tissues. Emerg Infect Dis 16(7):1151–1154
Jacobs JG, Langeveld JP et al (2007) Molecular discrimination of atypical bovine spongiform encephalopathy strains from a geographical region spanning a wide area in Europe. J Clin Microbiol 45(6):1821–1829
Jeffrey M, Martin S et al (2006) Immunohistochemical features of PrP(d) accumulation in natural and experimental goat transmissible spongiform encephalopathies. J Comp Pathol 134(2–3):171–181

Kong Q, Zheng M et al (2008) Evaluation of the human transmission risk of an atypical bovine spongiform encephalopathy prion strain. J Virol 82(7):3697–3701

Konold T, Bone G et al (2012) Experimental H-type and L-type bovine spongiform encephalopathy in cattle: observation of two clinical syndromes and diagnostics challenges. BMC Vet Res 8(22)

Lasmezas CI, Deslys JP et al (1996) BSE transmission to macaques. Nature 381(6585):743–744

Lombardi G, Casalone C et al (2008) Intraspecies transmission of BASE induces clinical dullness and amyotrophic changes. PLoS Pathog 4(5):e1000075

Oelschlegel AM, Fallahi M et al (2012) The extended cell panel assay characterizes the relationship of prion strains RML, 79A, and 139A and reveals conversion of 139A to 79A-like prions in cell culture. J Virol 86(9):5297–5303

Okada H, Iwamaru Y et al (2011a) Properties of L-type bovine spongiform encephalopathy in intraspecies passages. Vet Pathol. Published on-line 11 November 2011 at vet.sagepub.com/content/early

Okada H, Iwamaru Y et al (2011b) Experimental H-type bovine spongiform encephalopathy characterized by plaques and glial- and stellate-type prion protein deposits. Vet Res 42(1):79

Ortiz-Pelaez A, Stevenson MA et al (2012) Case-control study of cases of bovine spongiform encephalopathy born after July 31, 1996 (BARB cases) in Great Britain. Vet Rec 170(15):389

Peden A, McCardle L et al (2010) Variant CJD infection in the spleen of a neurologically asymptomatic UK adult patient with haemophilia. Haemophilia 16(2):296–304

Polak MP, Zmudzinski JF et al (2008) Atypical status of bovine spongiform encephalopathy in Poland: a molecular typing study. Arch Virol 153(1):69–79

Prusiner SB (1997) Prion diseases and the BSE crisis. Science 278(5336):245–251

Suardi S, Vimercati C et al (2012) Infectivity in skeletal muscle of cattle with atypical bovine spongiform encephalopathy. PLoS One 7(2):e31449

van Keulen LJ, Schreuder BE et al (2000) Pathogenesis of natural scrapie in sheep. Arch Virol Suppl (16):57–71

Weissmann C (2012) Mutation and selection of prions. PLoS Pathog 8(3):e1002582

Wells GA, Scott AC et al (1987) A novel progressive spongiform encephalopathy in cattle. Vet Rec 121(18):419–420

Wells GAH, Hawkins SAC et al (1998) Preliminary observations on the pathogenesis of experimental bovine spongiform encephalopathy (BSE): an update. Vet Rec 142(5):103–106

Wells GA, Konold T et al (2007) Bovine spongiform encephalopathy: the effect of oral exposure dose on attack rate and incubation period in cattle. J Gen Virol 88(Pt 4):1363–1373

Wickner RB, Edskes HK et al (2009) Prion variants, species barriers, generation and propagation. J Biol 8(5):47

Wilesmith JW, Wells GA et al (1988) Bovine spongiform encephalopathy: epidemiological studies. Vet Rec 123(25):638–644

Will RG, Ironside JW et al (1996) A new variant of Creutzfeldt-Jakob disease in the UK. Lancet 347(9006):921–925

Pathogenic *Escherichia coli* and One Health Implications

Narelle Fegan and Kari S. Gobius

Abstract *Escherichia coli* are common inhabitants in the intestinal tracts of warm blooded animals where they generally cause no harm to the host, although there are some types of *E. coli* which are able to cause disease. The most significant of these are enterohaemorrhagic *E. coli* (EHEC) which can cause severe human disease that can result in death. EHEC have an animal reservoir, particularly cattle, and are considered to be an important zoonotic pathogen having significant impact for One Health. EHEC can be transmitted from animals into humans, either from consumption of foods made from these animals, or from contact with foods which may have become contaminated directly or indirectly from animal wastes. Increasingly, EHEC have also been associated with uncooked leafy green vegetables and sprouts. Several large outbreaks of *E. coli* have highlighted the importance for addressing these organisms in a One Health perspective.

Contents

N. Fegan (✉)
CSIRO Animal, Food and Health Sciences, 671 Sneydes Rd, Werribee, VIC 3030, Australia
e-mail: Narelle.Fegan@csiro.au

K. S. Gobius
CSIRO Animal, Food and Health Sciences, 39 Kessels Rd, Coopers Plains, QLD 4108, Australia
e-mail: Kari.Gobius@csiro.au

Current Topics in Microbiology and Immunology (2013) 366: 49–62
DOI: 10.1007/82_2012_261

Published Online: 24 August 2012

1 Introduction

Escherichia coli is an organism that is found in the intestines of warm blooded animals. In most situations *E. coli* exists without causing harm to the host in which it resides, but some types of *E. coli* are able to cause disease in animals while others cause disease in humans. *E. coli* is typically transmitted via the faecal oral route thereby making pathogenic types important food and water borne pathogens. Although most pathogenic *E. coli* are thought to be species specific, those *E. coli* which can cross from animals into humans to cause disease are of particular interest from a One Health perspective. The route by which these *E. coli* enter the human host from their animal reservoir can be straightforward, such as direct contact or consumption of foods made from animals, or more complicated involving not only the original animal reservoir, but other animals and environmental factors. *E. coli* is therefore an important pathogen in relation to One Health.

2 *Escherichia coli* as Pathogens of Humans and Animals

Most *E. coli* reside within the intestine without causing any disease or harm to their host. Some types of *E. coli* are pathogenic to their hosts, either human or animal, and are capable of causing gastrointestinal illness, such as diarrhoea; or extraintestinal diseases, such as urinary tract infections, meningitis and septiceamia (Sousa 2006). In humans, the greatest disease burden from *E. coli* results from intestinal infections with disease severity ranging from self limiting diarrhoea through bloody diarrhoea to severe complications that may lead to death. *E. coli* is also a pathogen of veterinary significance as it causes disease in livestock leading to economic losses, examples include mastitis in cattle and pigs (Gerjets and Kemper 2009; Shpigel et al. 2008) and colibacillosis in poultry (Olsen et al. 2011). *E. coli* can also cause disease in companion animals such as dogs and cats (Mainil 2002).

There are many different types of pathogenic *E. coli* which are characterised based on their virulence determinants, clinical symptoms and serology (Wasteson 2002), the most important of which are discussed below. With the exception of enterohaemorrhagic *E. coli* (EHEC), most pathogenic *E. coli* are currently not considered to be major zoonotic pathogens due to their host specificity. However, *E. coli* has the ability to exchange genetic material and there is the potential for new pathogenic types to emerge which may be able to cross species barriers and cause disease in humans. These pathogenic types are briefly mentioned in the following section as they pose a risk in relation to being emerging zoonoses in the future.

The most common extraintestainal infections caused by *E. coli* are those of the urinary tract which are associated with uropathogenic *E. coli* (UPEC). Although most UPEC are thought to come from the intestinal tract of their host, there is some emerging evidence that animals may be a potential reservoir for these pathogens (Belanger et al. 2011), although it is still unclear as to the contribution of animals as a significant reservoir for extraintestinal pathogenic *E. coli.*

There are several important groups of pathogenic *E. coli* that cause gastrointestinal illness in both humans and animals, although there is little evidence for zoonotic transfer in these groups. This includes the enterotoxigenic *E. coli* (ETEC), the major cause of travellers' diarrhoea, which produce heat labile and heat stable toxins and cause a watery diarrhoea. ETEC can cause disease in humans and in newborn animals, but the colonisation factors they use to adhere to the host intestine are species specific with human isolates producing different types to those ETEC infecting animals (Nataro and Kaper 1998). EPEC are an important cause of infant diarrhoea in developing countries and have also been associated with disease in animals. One of the major features of EPEC is their ability to cause attaching and effacing (A/E) lesions on intestinal epithelial cells. This results from intimate adherence of the bacterium to the microvilli and subsequent changes in the epithelial cell cytoskeleton resulting in the formation of a pedestal structure beneath the site of bacterial attachment. The genes responsible for A/E are encoded on a pathogenicity island known as the locus of enterocyte effacement (LEE) (Nataro and Kaper 1998). Attachment mechanisms, including a protein encoded on LEE called intimin, and fimbriae, such as the bundle forming pilus are important in EPEC pathogenesis and are also likely to play a role in host specificity (Mundy et al. 2007; Bardiau et al. 2010).

Some pathogenic *E. coli* groups appear to be specific to humans only. Enteroaggregative *E. coli* (EAEC) can cause both persistent diarrhoea lasting for more than 14 days, and acute diarrhoea which has also been associated with travelling to developing countries. This group of *E. coli* are poorly defined and comprise a very heterogeneous group of *E. coli.* The uniting feature of EAEC is the formation of a characteristic 'stacked-brick' adherence pattern referred to as aggregative adherence (AA) due to the presence of adherence factors (Huang et al. 2006). There is no obvious animal reservoir for EAEC, although very little is known about the ecology of this type of *E. coli* outside of the human host. The enteroinvasive *E. coli* (EIEC), which are closely related to *Shigella* spp. and cause similar disease, appear to be a human specific pathogenic group of *E. coli* and have not been isolated from any animal sources (Beutin 1999).

The group of pathogenic *E. coli* which causes the most severe disease in humans do have an animal reservoir, these are the EHEC which are generally defined by the presence of LEE and the ability to produce a group of toxins known as Shiga toxins (Stx) (Nataro and Kaper 1998). This makes EHEC the most important group of *E. coli* from a One Health perspective and they will be the major focus of this chapter. EHEC can cause severe human disease that can result in death, one of the major reservoirs of EHEC is healthy adult ruminant animals, although some strains of EHEC can cause diarrhoea in young animals (Hornitzky et al. 2005). Foods made from animals which are carrying EHEC may become contaminated and large outbreaks of disease have resulted. For these reasons, *E. coli* has been prioritised by the World Health Organisation for the development of future standards for animal production and food safety (Knight-Jones et al. 2010). The best known of the EHEC serotypes is *E. coli* O157:H7, but many other serotypes are also capable of causing severe disease in humans (Karmali 2005).

3 Human Disease, Pathogenesis and Treatment

EHEC cause disease in humans that can range from asymptomatic carriage, through diarrhoea and bloody diarrhoea (haemorrhagic colitis, HC) to the severe disease haemolytic uraemic syndrome (HUS) which has the potential to be fatal (Karmali 2005). Symptoms of infection with EHEC may initially include abdominal cramps and diarrhoea which can occur in 1–4 days after infection, HC may also occur and HUS will develop in about 10–25 % of cases (Nataro and Kaper 1998; Karmali 1989). HUS may occur in 5–13 days after diarrhoea and comprises acute renal failure leading to reduced urine excretion, microangiopathic haemolytic anaemia (narrowing of small blood vessels causing fragmentation of erythrocytes, haemolysis and anaemia) and thromobocytopaenia (low platelet levels) (Tarr et al. 2005). In most cases patients will fully recover from HUS, but severe kidney damage and other complications will occur in about 12–30 % of cases and about 5 % will be fatal (Nataro and Kaper 1998). Additional complications may occur and include myocardial dysfunction, pancreatitis, hepatitis, pulmonary edema and neurological impairment (Tarr et al. 2005). EHEC infections are the leading cause of acute renal failure in children (Karmali et al. 2010) and together with the potential to cause fatalities, this group of pathogenic *E. coli* have become a high priority for control.

The ability of EHEC to cause disease results from the presence of two major virulence factors, the production of Stx which elicit the most severe outcomes associated with EHEC infection, and LEE, enabling the bacterium to colonise the intestine through the A/E cytopathology (Nataro and Kaper 1998). Many *E. coli* other than EHEC carry Stx, these are broadly known as Shiga toxigenic *E. coli* (STEC) and are commonly found in animals, though not all are not believed to cause severe human disease (Karmali 2005). There are two types of Stx, Stx1 and Stx2, the former is closely related to the toxin produced by *Shigella dysenteriae* (Pennington 2010). Stx are encoded on bacteriophages integrated into the bacterial genome and are produced by the bacteria in the intestine. The toxins are able to translocate across the epithelium and enter the bloodstream. Stx binds specifically to glycosphingolipid globotriacylceramide (Gb3) which is present on the surfaces of cells in the human kidney (Pennington 2010). Once bound to Gb3, Stx is internalized by the cell and causes protein synthesis inhibition and eventually cell death (Pennington 2010). LEE facilitates attachment of EHEC to the intestinal epithelium, but it is not critical for HUS to occur and some serotypes of EHEC utilise other adherence mechanisms to colonise the intestine while producing Stx (Doughty et al. 2002; Wu et al. 2010). A greater understanding of the full complement of virulence factors required by STEC and EHEC to illicit disease in humans is therefore an important area of further research.

The treatments available for EHEC infections are limited and many are still under development. Most approaches rely on supportive care such as maintaining fluids and electrolytes (Goldwater and Bettelheim 2012). The production of toxins by EHEC may occur shortly after infection and will continue to circulate through the body after the EHEC has been cleared from the intestine (Nataro and Kaper 1998). The use of antibiotics is generally avoided as they can increase the release of the Stx from bacteria (Wong et al. 2000). Various approaches have been used to try and limit the action of the Stx during infection and prevent progression of the disease to HUS. These include using neutralising Stx specific antibodies and other types of toxin binders and neutralisers (Goldwater and Bettelheim 2012). Vaccine strategies have been developed with variable success in animal models but there is still some way to go before they are available for human use (Goldwater and Bettelheim 2012). Prevention of initial infection is clearly the most effective way to prevent serious illness developing.

4 Disease Outbreaks and Links to Animals

EHEC have a low infectious dose with disease occurring from ingestion of less than 10 cells (Hara-Kudo and Takatori 2011); therefore, exposure to only small amounts of contamination can cause a health risk. Many cases of EHEC infection are sporadic with no apparent food or animal source. Person-to-person transmission can lead to outbreaks, particularly amongst young children at child care centres (Gilbert et al. 2008; Raffaelli et al. 2007), but more often outbreaks are associated with foods, particularly those derived from animal sources. EHEC were responsible for 84 outbreaks in England and Wales between 1992 and 2008 resulting in 1,168 affected individuals of which 286 were hospitalised and 12 deaths occurred (Gormley et al. 2011). Of the 44 outbreaks for which a food vehicle was identified, red meat was associated with 37 % and milk and milk products with 30 % indicating a strong association between EHEC outbreaks and foods derived from cattle and animal sources (Gormley et al. 2011).

Direct contact with animals and their immediate environments, often at petting zoos, farms, rodeos or fairs, has been the source of several EHEC outbreaks (Stirling et al. 2008; Steinmuller et al. 2006; Lanier et al. 2011). Fruit juices have been implicated in EHEC outbreaks where the fruits have come into contact with contaminated sewage, manures or soils (Vojdani et al. 2008). Unpasteurised apple juice caused an outbreak of *E. coli* O157:H7 infection which affected 70 people from the western United States and British Columbia in 1996 (Cody et al. 1999). The orchards supplying some of the fruit hosted a deer population which were shedding *E. coli* O157:H7 and the investigation into the outbreak concluded that dropped apples were most likely contaminated from contact with the ground and deer faeces (Cody et al. 1999). Many other fresh produce outbreaks have occurred with contamination thought to result from contact between the plants and manures, sewage, irrigation water and runoff (Beuchat 2006). Runoff from farms can also contaminate drinking water sources and result in outbreaks if treatment systems are inadequate. A large outbreak caused by both *Campylobacter* and *E. coli* O157:H7 occurred in Walkerton, Canada during 2000. The outbreak involved over 2,300 cases of gastroenteritis (in a town of 4,800 residents) with 27 cases of HUS and seven deaths (Hrudey et al. 2003). Heavy rainfall resulted in runoff from farms contaminating the ground water in wells that fed the municipal water supply, at the same time, chlorination equipment failed at the water treatment plant leading to contamination of the water supply (Danon-Schaffer 2001). These outbreaks highlight the complex chain of events and transmission pathways that often lead to human disease.

5 Sources and Transmission Routes

EHEC of serotype O157:H7 have been isolated from many ruminant animals, including sheep, goats, bison, deer and water buffalo, but the most important reservoir has been identified as cattle (Ferens and Hovde 2011). *E. coli* O157:H7 has also been isolated on occasion from a range of other animals including pigs, dogs, rats, rabbits, horses, amphibians, fish, various kinds of birds and insects (including flies and dung beetles). These animals are not considered to be a significant reservoir of this pathogen but they may play a minor role in transmission between animals, humans and the environment (Garcia et al. 2010; Ferens and Hovde 2011). Even though cattle and ruminants are considered to be the most important zoonotic source of *E. coli* O157:H7, as yet unidentified animal reservoirs may also exist which serve to act as long-term sources of cattle infection (Garcia et al. 2010). EHEC serotypes other than *E. coli* O157:H7 (such as O111 and O26) have been isolated from a variety of animals (Bettelheim 2007), including young animals with diarrhoea (Hornitzky et al. 2005; Jenkins et al. 2008; Badouei et al. 2010).

EHEC, whose primary reservoir is cattle, are mostly found in the intestine and can be shed in their faeces, but they are also commonly found on the hides and in the mouths of cattle (Fegan et al. 2005; Keen and Elder 2002). *E. coli* O157:H7 are thought to selectively colonise cattle at the distal region of the rectum near the recto-anal junction (Naylor et al. 2003) which is mediated by factors associated with LEE and a plasmid carried by *E. coli* O157:H7 (Sheng et al. 2006; Naylor et al. 2005). Such specific tissue tropism has not been reported for other EHEC serotypes which are found throughout the intestinal tract (Aktan et al. 2007). The prevalence of EHEC in animals ranges widely with surveys finding anywhere between none and the majority of animals shedding these pathogens (Barlow and Mellor 2010; Rhoades et al. 2009; Masana et al. 2010; Kobayashi et al. 2009;

Hussein 2007). Most of what is understood about animal carriage of EHEC comes from work focussing on *E. coli* O157:H7. Shedding of *E. coli* O157:H7 by animals is variable and often dependant on the age of the animal (higher in animals between 2 months and 2 years old) and the weather (increased shedding observed in the warmer months in some countries) (Garcia et al. 2010). Shedding of *E. coli* O157:H7 in animal faeces is generally intermittent with the majority of animals shedding for only a short period of time, although some animals may become persistent shedders (Robinson et al. 2004). The majority of animals shed low numbers of *E. coli* O157:H7 in their faeces (<1,000 or 10,000 cfu/g of faeces), but a few animals shed higher numbers and these have been termed high-level carriers or super-shedders (Low et al. 2005; Omisakin et al. 2003). A few super-shedding animals can be responsible for contributing >96 % of the total *E. coli* O157:H7 load (Omisakin et al. 2003) and represent the greatest risk for transmission within herds (Cobbold et al. 2007; Matthews et al. 2006) and contamination of carcases during processing (Fegan et al. 2005, 2009). Super-shedding is also believed to occur for other EHEC serotypes (Menrath et al. 2010). The factors influencing super-shedding are currently unknown, but targeting such animals will be important in efforts to reduce the risk of EHEC infections in humans (Chase-Topping et al. 2008).

Transmission routes of EHEC between animals and humans can be direct, such as contact between animals and humans, or can be far more complicated. Foods made from animals harbouring EHEC, such as meat and dairy products, may become contaminated during production and processing and lead to human disease if consumed. *E. coli* shed in the faeces of cattle can contaminate farm environments as manure and waste materials can contain EHEC (Fremaux et al. 2008). *E. coli* can survive for long periods of time (several months or longer) in soils, depending on the composition of the soil and also in manures, although proper composting or waste treatment can kill *E. coli* O157:H7 (Fremaux et al. 2008; Ferens and Hovde 2011). Rainfall events can assist in the spread of *E. coli* from farms to nearby water sources and contaminated water may result in illness if it is used to irrigate fresh produce (Hilborn et al. 1999) or used for swimming or drinking (Olsen et al. 2002; Centres for Disease Control and Prevention 1996; Hrudey et al. 2003). Understanding the mechanisms involved in EHEC transmission between animals, the environment and humans, and developing controls to limit this transmission will be of great importance to reduce the risk of this pathogen causing human illness.

6 Controlling EHEC

There are few effective treatments for EHEC infection in humans. Prevention of ingestion of the organism is therefore the most effective control and focussing on reducing the pathogen load within food production systems and the environment will be the most important aspect for prevention of disease (Garcia et al. 2010;

Khanna et al. 2008). To be effective in reducing the risk of human EHEC infection, controls must be focussed on all aspects enabling the transmission of these organisms including interventions aimed at animals, environments, the food chain and humans. Pre-harvest control in animal populations has used a variety of approaches to reduce EHEC shedding. These include the use of vaccines against *E. coli* O157:H7, lytic bacteriophages specifically targeting *E. coli* O157:H7, probiotics and direct fed microbials, diet manipulation and feed additives. These controls go some way to reducing the prevalence of EHEC in animal populations but do not completely eliminate the pathogen (Berry and Wells 2010). The costs, frequency of application, efficacy and regulatory requirements will all impact on the adoption of such controls into the future. Reducing animal shedding of EHEC will decrease the incidence of these organisms in animal-based food products and in the environment thereby limiting the opportunity for transmission into humans. Appropriate composting and preventing runoff of manures and wastes will reduce environmental contamination and that of fresh produce and water supplies. Post-harvest controls during food production, such as pasteurisation of milk for use in dairy products, hide and carcase decontamination in meat production, also lead to the reduction of EHEC in food products and reduce the risk of human infection (Berry and Wells 2010; Oliver et al. 2009). Good manufacturing process and ensuring hygiene throughout at all stages of food production (from farm to retail and consumption) are critical for reducing human infection with EHEC.

7 Case Studies

The significance of the One Health paradigm to EHEC will now be illustrated through two case studies. The complexities of human activities for plant and animal food production, globalised food trade, use of antibiotics for the control of human and animal illness, alongside microbial ecology, all interact to demonstrate the sensitive interrelationships which can profoundly influence One Heath.

Case study 1. During 2006 a large multistate foodborne outbreak of *E. coli* O157 occurred in the United States. This outbreak was distinctive due to the severity of disease symptoms and the resultant high rates of both hospitalisation (50 %) and HUS (10 %) (Manning et al. 2008). While the most common food vehicle for EHEC infection in the US is undercooked beef hamburger, in this outbreak, bagged baby spinach leaves were found to be the food source (CDC 2006). The contaminated spinach was traced to a single day of production at one processing plant and fields located on four spinach farms on the central California coast. Further investigations determined that the faeces of feral swine likely contaminated the spinach following transmission (to the swine) of *E. coli* O157 from cattle located in the same geographical region, but not present on the immediate fresh produce farms. Surface water was also considered as a potential pathway for the transmission of the O157 from cattle to swine (Jay et al. 2007). Subsequent molecular and genomic analysis of the pathogenic outbreak strain revealed a distinct genome organisation, placing the strain in a 'hypervirulence' clade (Manning et al. 2008), and integration of a Stx2-prophage genome at a novel site in the bacterial chromosome (Kulasekara et al. 2009). This novel Stx2-prophage directs the production of elevated levels of Stx2 toxin which may account for the hypervirulence of such *E. coli* O157 strains (Neupane et al. 2011).

Case study 2. The application of bacterial genome sequence-based genotyping combined with product trace-back enabled identification of a particularly unusual food source and a rare HUS-causing pathogenic *E. coli* associated with a massive outbreak in Europe during 2011. *E. coli* serotype O104:H4 infection resulting from ingestion of salad sprouts from imported fenugreek seeds led to a total of 3,842 cases, and 855 cases of HUS leading to 35 deaths with another 18 deaths from nonHUS complications associated with infection (Muniesa et al. 2012). While the outbreak strain was never isolated from fenugreek seeds, molecular typing methods were applied to strains from separate French and German outbreaks which indicated a common causative pathogen and facilitated identification of a common food source. Rapid genome sequencing revealed that the infectious agent was not *E. coli* O157, but rather a member of the EAEC pathotype which had acquired a Stx2-encoding prophage through the process of bacteriophage transduction (Rohde et al. 2011). The genome sequence also indicated that no fewer than 10 different antibiotic resistance genes had been acquired by the outbreak strain. Clearly, the specific combination of EAEC pathogenesis mechanisms, augmented by Stx production, constituted a particularly powerful virulence arsenal (Rohde et al. 2011).

The case studies presented above demonstrate selected elements of the One Health web. If viewed through a narrow lens, the different nodes of the web may appear separate and unconnected. However, if viewed through a wide-angle lens, the web of interconnection becomes abundantly clear. In both cases, uncooked vegetable foods (normally considered healthy and nutritious) were the vehicle carrying infectious *E. coli* to the human gastrointestinal tract. Since EAEC O104:H4 is known to be restricted to human hosts (Kuijper et al. 2011) (unlike the zoonotic EHEC O157), it is very likely that O104:H4 contamination of fenugreek seeds resulted either directly or indirectly from human faecal contamination. The plausibility of potential contamination by human faeces is supported by the resistance to multiple antibiotics of the outbreak strain. Such antibiotic resistance is selected by recurrent antibiotic use, which suggests that the organism may have emerged in humans suffering diarrhoea who were then treated with antibiotics for such illness. Since some classes of antibiotic are known to promote the rapid production of Stx bacteriophages by their host *E. coli* cells (Bielaszewska et al. 2012), it is also feasible that transfer of the Stx2-encoding prophage to the O104:H4 EAEC strain was induced by human antibiotic use.

Our emerging recognition of the significance of One Health will now foster further effort to generate deeper knowledge of its intricate relationships. This knowledge will then support endeavours to steward the finely balanced interdependency of human, animal, plant and environmental health.

References

Aktan I, La Ragione RM, Woodward MJ (2007) Colonization, persistence, and tissue tropism of *Escherichia coli* O26 in conventionally reared weaned lambs. Appl Environ Microbiol 73(3):691–698

Badouei MA, Salehi TZ, Khorasgani MR, Tadjbakhsh H, Brujeni GN (2010) Occurrence and characterisation of enterohaemorrhagic *Escherichia coli* isolates from diarrhoeic calves. Comp Clin Pathol 19(3):295–300. doi:10.1007/s00580-009-0873-0

Bardiau M, Szalo M, Mainil JG (2010) Initial adherence of EPEC, EHEC and VTEC to host cells. Vet Res 41(5):57. doi:10.1051/vetres/2010029

Barlow RS, Mellor GE (2010) Prevalence of enterohemorrhagic *Escherichia coli* serotypes in Australian beef cattle. Foodborne Pathog Dis 7(10):1239–1245

Belanger L, Garenaux A, Harel J, Boulianne M, Nadeau E, Dozois CM (2011) *Escherichia coli* from animal reservoirs as a potential source of human extraintestinal pathogenic *E. coli*. FEMS Immunol Med Microbiol 62(1):1–10. doi:10.1111/j.1574-695X.2011.00797.x

Berry E, Wells J (2010) *Escherichia coli* O157:H7 recent advances in research on occurrence, transmission, and control in cattle and the production environment. Adv Food Nutr Res 60:67–117

Bettelheim KA (2007) The non-O157 Shiga-toxigenic (verocytotoxigenic) *Escherichia coli*; under-rated pathogens. Crit Rev Microbiol 33(1):67–87

Beuchat LR (2006) Vectors and conditions for preharvest contamination of fruits and vegetables with pathogens capable of causing enteric diseases. Br Food J 108(1):38–53. doi:10.1108/00070700610637625

Beutin L (1999) *Escherichia coli* as a pathogen in dogs and cats. Vet Res 30(2–3):285–298

Bielaszewska M, Idelevich EA, Zhang W, Bauwens A, Schaumburg F, Mellmann A, Peters G, Karch H (2012) Effects of antibiotics on Shiga toxin 2 production and bacteriophage induction by epidemic *Escherichia coli* O104:H4 strain. Antimicrob Agents Chemother 56(6):3277–3282. doi:10.1128/AAC.06315-11, AAC.06315-11 [pii]

CDC (2006) Ongoing multistate outbreak of *Escherichia coli* serotype O157:H7 infections associated with consumption of fresh spinach—United States, Sep 2006. MMWR Morb Mortal Wkly Rep 55(38):1045–1046. mm5538a4 [pii]

Centers for Disease Control and Prevention (1996) Lake-associated outbreak of *Escherichia coli* O157:H7—Illinois, 1995. MMWR Morb Mortal Wkly Rep 45(21):437–439

Chase-Topping M, Gally D, Low C, Matthews L, Woolhouse M (2008) Super-shedding and the link between human infection and livestock carriage of *Escherichia coli* O157. Nat Rev 6(12):904–912

Cobbold RN, Hancock DD, Rice DH, Berg J, Stilborn R, Hovde CJ, Besser TE (2007) Rectoanal junction colonization of feedlot cattle by *Escherichia coli* O157:H7 and its association with supershedders and excretion dynamics. Appl Environ Microbiol 73(5):1563–1568

Cody SH, Glynn MK, Farrar JA, Cairns KL, Griffin PM, Kobayashi J, Fyfe M, Hoffman R, King AS, Lewis JH, Swaminathan B, Bryant RG, Vugia DJ (1999) An outbreak of *Escherichia coli* O157: H7 infection from unpasteurized commercial apple juice. Ann Intern Med 130(3): 202–209

Danon-Schaffer MN (2001) Walkerton's contaminated water supply system: a forensic approach to identifying the source. Environ Forensics 2(3):197–200. doi:10.1006/enfo.2001.0054

Doughty S, Sloan J, Bennett-Wood V, Robertson M, Robins-Browne RM, Hartland EL (2002) Identification of a novel fimbrial gene cluster related to long polar fimbriae in locus of enterocyte effacement-negative strains of enterohemorrhagic *Escherichia coli*. Infect Immun 70(12):6761–6769. doi:10.1128/iai.70.12.6761-6769.2002

Fegan N, Higgs G, Vanderlinde P, Desmarchelier P (2005) An investigation of *Escherichia coli* O157 contamination of cattle during slaughter at an abattoir. J Food Prot 68(3):451–457

Fegan N, Higgs G, Duffy LL, Barlow RS (2009) The effects of transport and lairage on counts of *Escherichia coli* O157 in the feces and on the hides of individual cattle. Foodborne Pathog Dis 6(9):1113–1120. doi:10.1089/fpd.2009.0338

Ferens WA, Hovde CJ (2011) *Escherichia coli* O157:H7: animal reservoir and sources of human infection. Foodborne Pathog Dis 8(4):465–487. doi:10.1089/fpd.2010.0673

Fremaux B, Prigent-Combaret C, Vernozy-Rozand C (2008) Long-term survival of Shiga toxin-producing *Escherichia coli* in cattle effluents and environment: an updated review. Vet Microbiol 132(1–2):1–18

Garcia A, Fox JG, Besser TE (2010) Zoonotic enterohemorrhagic *Escherichia coli*: a one health perspective. ILAR J 51(3):221–232

Gerjets I, Kemper N (2009) Coliform mastitis in sows: a review. J Swine Health Prod 17(2):97–105

Gilbert M, Monk C, Wang HL, Diplock K, Landry L (2008) Screening policies for daycare attendees—lessons learned from an outbreak of *E.coli* O157: H7 in a daycare in Waterloo, Ontario. Can J Public Health Rev Can Sante Publ 99(4):281–285

Goldwater PN, Bettelheim KA (2012) Treatment of enterohemorrhagic *Escherichia coli* (EHEC) infection and hemolytic uremic syndrome (HUS). BMC Med 10:12. doi:10.1186/1741-7015-10-12

Gormley FJ, Little CL, Rawal N, Gillespie IA, Lebaigue S, Adak GK (2011) A 17-year review of foodborne outbreaks: describing the continuing decline in England and Wales (1992–2008). Epidemiol Infect 139(5):688–699. doi:10.1017/s0950268810001858

Hara-Kudo Y, Takatori K (2011) Contamination level and ingestion dose of foodborne pathogens associated with infections. Epidemiol Infect 139(10):1505–1510. doi:10.1017/s095026881000292x

Hilborn ED, Mermin JH, Mshar PA, Hadler JL, Voetsch A, Wojtkunski C, Swartz M, Mshar R, Lambert-Fair MA, Farrar JA, Glynn MK, Slutsker L (1999) A multistate outbreak of *Escherichia coli* O157:H7 infections associated with consumption of mesclun lettuce. Arch Intern Med 159(15):1758–1764

Hornitzky MA, Mercieca K, Bettelheim KA, Djordjevic SP (2005) Bovine feces from animals with gastrointestinal infections are a source of serologically diverse atypical enteropathogenic *Escherichia coli* and Shiga toxin-producing *E. coli* strains that commonly possess intimin. Appl Environ Microbiol 71(7):3405–3412

Hrudey SE, Payment P, Huck PM, Gillham RW, Hrudey EJ (2003) A fatal waterborne disease epidemic in Walkerton, Ontario: comparison with other waterborne outbreaks in the developed world. Water Sci Technol 47(3):7–14

Huang DB, Mohanty A, DuPont HL, Okhuysen PC, Chiang T (2006) A review of an emerging enteric pathogen: enteroaggregative *Escherichia coli*. J Med Microbiol 55(10):1303–1311. doi:10.1099/jmm.0.46674-0

Hussein HS (2007) Prevalence and pathogenicity of Shiga toxin-producing *Escherichia coli* in beef cattle and their products. J Anim Sci 85:E63–E72. doi:10.2527/jas.2006-421

Jay MT, Cooley M, Carychao D, Wiscomb GW, Sweitzer RA, Crawford-Miksza L, Farrar JA, Lau DK, O'Connell J, Millington A, Asmundson RV, Atwill ER, Mandrell RE (2007) *Escherichia coli* O157: H7 in feral swine near spinach fields and cattle, central California coast. Emerg Infect Dis 13(12):1908–1911

Jenkins C, Evans J, Chart H, Willshaw GA, Frankel G (2008) *Escherichia coli* serogroup O26—a new look at an old adversary. J Appl Microbiol 104(1):14–25. doi:10.1111/j.1365-2672.2007.03465.x

Karmali MA (1989) Infection by verocytotoxin-producing *Escherichia coli*. Clin Microbiol Rev 2(1):15–38

Karmali MA (2005) Use of comparative genomics as a tool to assess the clinical and public health significance of emerging shiga toxin-producing *Escherichia coli* serotypes. Meat Sci 71(1):62–71

Karmali MA, Gannon V, Sargeant JM (2010) Verocytotoxin-producing *Escherichia coli* (VTEC). Vet Microbiol 140(3–4):360–370. doi:10.1016/j.vetmic.2009.04.011

Keen JE, Elder RO (2002) Isolation of Shiga-toxigenic *Escherichia coli* O157 from hide surfaces and the oral cavity of finished beef feedlot cattle. J Am Vet Med Assoc 220(6):756–763

Khanna R, Waechter L, Sargeant J, Clark WF, Garg AX (2008) Environmental prevention of human disease from verocytotoxin-producing *Escherichia coli*. Nephrol Dial Transplant 23(6):1819–1822

Knight-Jones TJD, Mylrea GE, Kahn S (2010) Animal production food safety: priority pathogens for standard setting by the World Organisation for Animal Health. Rev Sci Tech Off Int Epizoot 29(3):523–535

Kobayashi H, Kanazaki M, Ogawa T, Iyoda S, Hara-Kudo Y (2009) Changing prevalence of O-serogroups and antimicrobial susceptibility among STEC strains isolated from healthy dairy cows over a decade in Japan between 1998 and 2007. J Vet Med Sci 71(3):363–366

Kuijper EJ, Soonawala D, Vermont C, van Dissel JT (2011) Household transmission of haemolytic uraemic syndrome associated with Escherichia coli O104:H4 in the Netherlands. Euro Surveill 16(25):7–9

Kulasekara BR, Jacobs M, Zhou Y, Wu ZN, Sims E, Saenphimmachak C, Rohmer L, Ritchie JM, Radey M, McKevitt M, Freeman TL, Hayden H, Haugen E, Gillett W, Fong C, Chang J, Beskhlebnaya V, Waldor MK, Samadpour M, Whittam TS, Kaul R, Brittnacher M, Miller SI (2009) Analysis of the genome of the *Escherichia coli* O157:H7 2006 spinach-associated outbreak isolate indicates candidate genes that may enhance virulence. Infect Immun 77(9):3713–3721. doi:10.1128/iai.00198-09

Lanier WA, Hall JM, Herlihy RK, Rolfs RT, Wagner JM, Smith LH, Hyytia-Trees EK (2011) Outbreak of Shiga-toxigenic *Escherichia coli* O157:H7 infections associated with rodeo attendance, Utah and Idaho, 2009. Foodborne Pathog Dis 8(10):1131–1133. doi:10.1089/fpd.2011.0884

Low JC, McKendrick IJ, McKechnie C, Fenlon D, Naylor SW, Currie C, Smith DG, Allison L, Gally DL (2005) Rectal carriage of enterohemorrhagic *Escherichia coli* O157 in slaughtered cattle. Appl Environ Microbiol 71(1):93–97

Mainil J (2002) Pathogenic *Escherichia coli* strains from dogs and cats: IV)—general discussion. Ann Med Vet 146 (4):219–224

Manning SD, Motiwala AS, Springman AC, Qi W, Lacher DW, Ouellette LM, Mladonicky JM, Somsel P, Rudrik JT, Dietrich SE, Zhang W, Swaminathan B, Alland D, Whittam TS (2008) Variation in virulence among clades of *Escherichia coli* O157:H7 associated with disease outbreaks. Proc Nat Acad Sci USA 105(12):4868–4873

Masana M, Leotta G, Del Castillo L, D'Astek B, Palladino P, Galli L, Vilacoba E, Carbonari C, Rodriguez H, Rivas M (2010) Prevalence, characterization, and genotypic analysis of *Escherichia coli* O157:H7/NM from selected beef exporting abattoirs of Argentina. J Food Prot 73(4):649–656

Matthews L, McKendrick IJ, Ternent H, Gunn GJ, Synge B, Woolhouse ME (2006) Super-shedding cattle and the transmission dynamics of *Escherichia coli* O157. Epidemiol Infect 134(1):131–142

Menrath A, Wieler L, Heidemanns K, Semmler T, Fruth A, Kemper N (2010) Shiga toxin producing *Escherichia coli:* identification of non-O157:H7-super-shedding cows and related risk factors. Gut Pathog 2(1):7

Mundy R, Schuller S, Girard F, Fairbrother JM, Phillips AD, Frankel G (2007) Functional studies of intimin in vivo and ex vivo: implications for host specificity and tissue tropism. Microbiology-(UK) 153:959–967. doi:10.1099/mic.0.2006/003467-0

Muniesa M, Hammerl JA, Hertwig S, Appel B, Brussow H (2012) Shiga toxin-producing *Escherichia coli* O104:H4: a new challenge for microbiology. Appl Environ Microbiol 78(12):4065–4073. doi:10.1128/aem.00217-12

Nataro JP, Kaper JB (1998) Diarrheagenic *Escherichia coli*. Clin Microbiol Rev 11(1):142–201

Naylor SW, Low JC, Besser TE, Mahajan A, Gunn GJ, Pearce MC, McKendrick IJ, Smith DG, Gally DL (2003) Lymphoid follicle-dense mucosa at the terminal rectum is the principal site of colonization of enterohemorrhagic *Escherichia coli* O157:H7 in the bovine host. Infect Immun 71(3):1505–1512

Naylor SW, Roe AJ, Nart P, Spears K, Smith DG, Low JC, Gally DL (2005) *Escherichia coli* O157: H7 forms attaching and effacing lesions at the terminal rectum of cattle and colonization requires the LEE4 operon. Microbiology 151(Pt 8):2773–2781. doi:10.1099/mic.0.28060-0, 151/8/2773 [pii]

Neupane M, Abu-Ali GS, Mitra A, Lacher DW, Manning SD, Riordan JT (2011) Shiga toxin 2 overexpression in *Escherichia coli* O157:H7 strains associated with severe human disease. Microb Pathog 51(6):466–470. doi:10.1016/j.micpath.2011.07.009

Oliver SP, Boor KJ, Murphy SC, Murinda SE (2009) Food safety hazards associated with consumption of raw milk. Foodborne Pathog Dis 6(7):793–806. doi:10.1089/fpd.2009.0302

Olsen RH, Stockholm NM, Permin A, Christensen JP, Christensen H, Bisgaard M (2011) Multi-locus sequence typing and plasmid profile characterization of avian pathogenic *Escherichia coli* associated with increased mortality in free-range layer flocks. Avian Pathol 40(5):437–444. doi:10.1080/03079457.2011.592822

Olsen SJ, Miller G, Kennedy M, Higgins C, Walford J, McKee G, Fox K, Bibb W, Mead P (2002) A waterborne outbreak of *Escherichia coli* O157:H7 infections and hemolytic uremic syndrome: implications for rural water systems. Emerging Infect Dis 8(4):370–375

Omisakin F, MacRae M, Ogden ID, Strachan NJ (2003) Concentration and prevalence of *Escherichia coli* O157 in cattle feces at slaughter. Appl Environ Microbiol 69(5):2444–2447

Pennington H (2010) *Escherichia coli* O157. Lancet 376(9750):1428–1435. doi:10.1016/S0140-6736(10)60963-4, S0140-6736(10)60963-4 [pii]

Raffaelli RM, Paladini M, Hanson H, Kornstein L, Agasan A, Slavinski S, Weiss D, Fennelly GJ, Flynn JT (2007) Child care-associated outbreak of *Escherichia coli* O157: H7 and hemolytic uremic syndrome. Pediatr Infect Dis J 26(10):951–953. doi:10.1097/INF.0b013e31812571f6

Rhoades J, Duffy G, Koutsoumanis K (2009) Prevalence and concentration of verocytotoxigenic *Escherichia coli*, *Salmonella enterica* and *Listeria monocytogenes* in the beef production chain: a review. Food Microbiol 26(4):357–376

Robinson SE, Wright EJ, Hart CA, Bennett M, French NP (2004) Intermittent and persistent shedding of *Escherichia coli* O157 in cohorts of naturally infected calves. J Appl Microbiol 97(5):1045–1053

Rohde H, Qin J, Cui Y, Li D, Loman NJ, Hentschke M, Chen W, Pu F, Peng Y, Li J, Xi F, Li S, Li Y, Zhang Z, Yang X, Zhao M, Wang P, Guan Y, Cen Z, Zhao X, Christner M, Kobbe R, Loos S, Oh J, Yang L, Danchin A, Gao GF, Song Y, Yang H, Wang J, Xu J, Pallen MJ, Aepfelbacher M, Yang R (2011) Open-source genomic analysis of Shiga-toxin-producing *E. coli* O104:H4. N Engl J Med 365(8):718–724. doi:10.1056/NEJMoa1107643

Sheng H, Lim JY, Knecht HJ, Li J, Hovde CJ (2006) Role of *Escherichia coli* O157:H7 virulence factors in colonization at the bovine terminal rectal mucosa. Infect Immun 74(8):4685–4693

Shpigel NY, Elazar S, Rosenshine I (2008) Mammary pathogenic *Escherichia coli*. Curr Opin Microbiol 11(1):60–65. doi:10.1016/j.mib.2008.01.004

Sousa CP (2006) The versatile strategies of *Escherichia coli* pathotypes: a mini review. J Venom Anim Tox incl Trop Dis 12(3):363–373

Steinmuller N, Demma L, Bender JB, Eidson M, Angulo FJ (2006) Outbreaks of enteric disease associated with animal contact: not just a foodborne problem anymore. Clin Infect Dis 43(12):1596–1602. doi:10.1086/509576

Stirling J, Griffith M, Dooley JSG, Goldsmith CE, Loughrey A, Lowery CJ, McClurg R, McCorry K, McDowell D, McMahon A, Millar BC, Rao J, Rooney PJ, Snelling WJ, Matsuda M, Moore JE (2008) Zoonoses associated with petting farms and open zoos. Vector Borne Zoonotic Dis 8(1):85–92. doi:10.1089/vbz.2006.0639

Tarr PI, Gordon CA, Chandler WL (2005) Shiga-toxin-producing *Escherichia coli* and haemolytic uraemic syndrome. Lancet 365(9464):1073–1086. doi:10.1016/S0140-6736(05)71144-2, S0140-6736(05)71144-2 [pii]

Vojdani JD, Beuchat LR, Tauxe RV (2008) Juice-associated outbreaks of human illness in the United States, 1995 through 2005. J Food Prot 71(2):356–364

Wasteson Y (2002) Zoonotic *Escherichia coli*. Acta Vet Scand 43:79–84

Wong CS, Jelacic S, Habeeb RL, Watkins SL, Tarr PI (2000) The risk of the hemolytic–uremic syndrome after antibiotic treatment of *Escherichia coli* O157:H7 infections. N Engl J Med 342(26):1930–1936. doi:10.1056/NEJM200006293422601

Wu YL, Hinenoya A, Taguchi T, Nagita A, Shima K, Tsukamoto T, Sugimoto N, Asakura M, Yamasaki S (2010) Distribution of virulence genes related to adhesins and toxins in Shiga toxin-producing *Escherichia coli* strains isolated from healthy cattle and diarrheal patients in Japan. J Vet Med Sci 72(5):589–597

Part II
National Plans for Developing a One Health Approach

FAO and the One Health Approach

Juan Lubroth

Abstract The Food and Agriculture Organization (FAO) of the United Nation's view on *One Health* is broad as it extends from human, animal—domestic and wildlife—and environmental health. Though the nidus of work originated within FAO's animal health service of the Agriculture and Consumer Protection Department, it is clearly an area of work that would include other departments such as Natural Resources Management and the Environment, Forestry, Fisheries and Aquaculture, Economic and Social Development, Legal Services, and communication. In terms of risk assessment and risk mitigation to health threats at the human–animal-ecosystem interface FAO works closely with its global partners, World Health Organisation and the World Organisation for animal health (the "Tripartite"). FAO's animal health service sees its work in *One Health* as contributing to all eight Millennium Development Goals, recognising the importance of animal health to human health, food safety, nutrition and food security, ameliorating poverty and hunger, natural resource management and partnerships. Some examples of FAO's operationalising *One Health* approaches or principles are introduced.

Contents

J. Lubroth (✉)
Animal Health Service, Food and Agriculture Organization, Viale delle Terme di Caracalla, 00153 Rome, Italy
e-mail: Juan.Lubroth@fao.org

Current Topics in Microbiology and Immunology (2013) 366: 65–72 65
DOI: 10.1007/82_2012_262

Published Online: 5 September 2012

1 Introduction

The increasing awareness of just how interconnected the systems are that intertwine the lives of animals, people and their environments—and each one of their healths—has come to be expressed as "One Health". As the human and animal global populations swell, the latter in large part to keep pace with growing food needs, human actions are placing enormous pressure on the viability and safety of the planet. Humans and their crops and livestock exact a toll on the environment; encroaching on pristine ecosystems for agriculture and development, exacerbating climate dynamics, exhausting soils and altering landscapes or contributing to further desertification. Greater urbanisation and sharp increases in international trade and transport—and the impressive speed with which pathogens can instantly become global, such as influenza, mean that in order to better protect global human health, greater care must be given to maintaining the health of ecosystems and animals. Two out of three new diseases in humans originate in the animal world—70 % of those from wildlife animals in particular (Jones et al. 2008). However, livestock raising also makes up 40 % of global agricultural GDP (FAO 2009). Livestock is especially critical in developing countries, where one billion of the poorest farmers depend on livestock for basic protein and as a way of making a living (Ashley et al. 1999).

One Health eludes any precise definition, making it hard to convey as a concept. FAO, like many, have resisted efforts to define it, recognising that part of its strength lies in the fact that it remains open to a range of disciplines that can contribute to global health. The importance of the term lies its interdisciplinarity—required for tackling complex challenges such as health, which is a continuum from the local to the global level. Although to date the effort and concept have been principally focused on infectious diseases, it is recognised that causation of disease states and the promotion of overall health is much broader. For safeguarding health, cohesion of collective efforts to improve nutrition and better understand the role of ecosystems in health requires exploration. For example by providing ecosystems services, biological diversity is maintained, thus benefiting improved biological resilience that provides a buffer against extreme events.

From an infectious disease perspective, FAO has established systems in prevention, detection, preparedness and response to better address threats at global and local levels (i.e., H5N1 avian influenza, foot-and-mouth disease, rabies, brucellosis, other high impact diseases). Critical to this work are FAO's collaborative agreements with the World Organisation for Animal Health (OIE) and the World Health Organization (WHO), another UN agency, as a tripartite working to promote *One Health* in common areas where synergies can be found (such as zoonotic agents, antimicrobial resistance, disease intelligence). The tripartite, individually and collectively, is involved in setting standards, improving governance and quality control mechanisms for health, as well as honing prevention efforts, developing early warning alerts and timely responses to disease events around the world and gaining better understanding of the drivers to disease emergence, maintenance and spread.

Zoonotic diseases and high impact animal diseases that have ruinous impacts on human health, livelihoods and commerce are given high priority.

FAO embraces the *One Health* approach by engaging the depth of resources and expertise it has as an organisation across the disciplines it has in-house: FAO has over 3,000 employees ranging from veterinarians, wildlife specialists, microbiologists, nutritionists, sociologists, lawyers, anthropologists, bioinformaticians, water engineers, crop economists, development communicators and extentionists, molecular biologists and emergency and relief responders. FAO is a *One Health* organisation.

Healthy animal production systems—terrestrial and aquatic; domestic and wild—converge with efforts to feed the world and manage natural resources sustainably, reducing the risk for disease outbreaks by implementing sound agricultural and development policies that improves people's food security and source of livelihood. FAO as an organisation, and in its *One Health* work, fosters community participation and gender equity in terms of resources, informations, goods, services and decision-making in rural and urban areas. The approach calls for increased public and private investments to strengthen existing agriculture and development systems and an enabling environment in which to do so. Getting wholesome food to the cities, with enormous peri-urban populations living in unhygienic conditions, remains a challenge for FAO and the *One Health* agenda.

While many issues are "One Health" issues, FAO's animal health service has prioritised actions and activities focusing on some ten priority infectious diseases because of their impact on food security, socio-economic importance, transboundary spread, food safety or economic security[1] and other cross-cutting issues such as antimicrobial resistance, information and surveillance systems between veterinary, wildlife and public health sectors, planning and preparedness. For FAO, the *One Health* priority issues extend beyond emerging and re-emerging infectious diseases, as some of the known and ancient diseases that have not been successfully addressed or curbed would benefit from an inter-disciplinary and multi-sectoral approach (i.e., tuberculosis or rabies). However, tackling single pathologies is likely not cost effective or efficient in the long run and a more encompassing approach would and could improve health across a production systems based on species or sector (i.e., small ruminant health, dairy sector). Having hygiene and biosecurity concepts practiced as if they were second nature can eliminate numerous disease threats; flock or herd health can include vaccination and treatment of the young stock for those high risk/high impact diseases.

It is recognised that an area of neglected importance is environmental health and ecosystem services, and further effort for outreach is required to fully capture the meaning of *One Health* and extend this beyond the mindset of an infectious zoonotic disease. Humans and other animals share the biosphere with all the other

[1] High impact diseases include: Peste des petits ruminants, foot-and-mouth disease, rift valley fever, brucellosis (especially *melitensis*), trypanosomosis, classical and African swine fever, sheep and goat pox, virulent avian influenza and Newcastle's disease, contagious caprine and bovine pleuropneumonias.

abiotic and biotic factors surrounding—including pathogens. An ecosystem that is healthy is resilient in terms of adapting or recovering from shocks. Such resilience also comes from the presence of high degree of biodiversity rather than a system of monocultures where pathogens may flourish unabated (Keesing et al. 2010; Garrett and Mundt 1999; Polis et al. 1997). The importance of ecosystem services include the transformation of a set of natural assets (soil, plants and animals, air and water, carbon sequestration and nitrogen fixation). If the natural assets decline, so do the benefits (Boyd and Banzhaf 2006). Thus FAO and others should look after and maintain these natural assets from which greater returns will come. Some of the examples of ecosystem services that come from nature include: pollination, regulation of climate, maintenance and provision of genetic resources, maintenance and regeneration of habitats, shade and shelter, prevention of soil erosion and maintenance of soil fertility, waterways and water filtration and waste absorption and breakdown.

Millennium Development Goals (United Nations 2012). The developing FAO contribution to *One Health* requires effective response mechanisms to redress poverty, health, hunger and natural resource management. This implies that a broad health management approach requires development to reflect that health security cannot be viewed in isolation, but includes aspects related to animal production, marketing, socio-economics, governance, poverty and requirements for targeted investments for development. In much of the developing world, food and health security are interdependent, as both are critical to the livelihoods of the more vulnerable groups of society. Thus the *One Health* approach and the importance of animals and animal products play an important role in contributing to the Millennium Development Goals (MDGs).

- MDG 1: Eradicate extreme poverty and hunger
  - This is one of the principal pillars of the FAO's mission. Livestock provide a highly nutritious source of food, provide draught power to till fields for crop production, to reap the harvest and transport it to market. Livestock also provide manure as nutrients for soils, and animal fibres and skins provide clothing, while other animal products can be used for shelter and housing material as well as adhesives. Livestock raising can provide a route out of poverty because of their economic value which can be moved in times of flooding, drought or civil unrest. Wildlife provide a valuable source of quality nutritions to millions of people.
- MDG 2: Achieve universal primary education
  - Quality nutrition in vulnerable households enables children to reach their full potential in school, having the energy, protein and micronutrients needed for their cognitive development.
- MDG 3: Promote gender equality and empower women
  - This is an important aspect in animal health and nutrition, since most often it is the women of a household that feed and care for the animals in smallholder

or family setting. Thus, improving animal health and production is only possible through women and by empowering women.

- MDG 4: Reduce child mortality
  - The improved incomes and reliability of incomes for poor households means that when an unexpected medical expense becomes urgent, those families can cope. The improved access to nutrition available from milk, dairy, eggs and meat also means that growing children have stronger immune systems to fend off illness and disease, including food and/or water-borne infections. In some nomadic societies, livestock are the sole source of milk and key to the survival of children under five.
- MDG 5: Improve maternal health
  - Maternal health depends on the micronutrients, vitamins and minerals often found in needed quantities throughout pregnancy and maternity, livestock and other animal products are often the unique source of these via milk, dairy, eggs and meat.
- MDG 6: Combat HIV/AIDS, malaria and other diseases
  - As with MDG 4 above, the body's ability to resist and fend off illness and disease will in part be determined by the individual's state of health when confronted with an infection. From an infectious disease viewpoint, many would argue that *One Health* relates principally (if not solely) to zoonotic diseases and other health threats at the human–animal interface. FAO broadens this view and considers that other animal diseases, while not zoonotic, can and do affect human health, nutrition and livelihoods.
- MDG 7: Ensure environmental sustainability
  - An additional FAO pillar is that of natural resource management and proper stewardship of the environment. This MDG captures the importance of having sound agricultural and land use policies, resilient ecosystems services and efficiencies in production parameters whereby the negative impact of food animals are minimised and conservation efforts are improved or at least maintained. In addition, the natural services (such as pollination, air and water quality) provided by nature are preserved and made efficient use of.
- Lastly, MDG 8: Develop a global partnership for development–recognises that no one group, country or agency can accomplish this alone, whether local or global.

FAO's Action Plan (Food and Agriculture Organisation 2011b). An Action Plan was endorsed by the FAO's Programme Committee in March 2011, which recognised FAO's comparative advantage in taking a broad, multidisciplinary approach and building on investments and lessons learned from the crisis unleashed by the H5N1 highly pathogenic avian influenza outbreaks that began in late 2003 in southeast Asia and peaked in 2006 with over 60 countries reporting its introduction. This Action Plan recognises that cooperation and collaboration with national governments, sub-regional, regional and global organisations and the engagement of donor development agencies are required.

The strategic vision guiding the Action Plan is a world in which risks to animal and animal-related human health due to a wide range of high impact zoonotic and non-zoonotic diseases, and their associated affect on food security, livelihoods, trade and economic development are minimised through improved upstream prevention measures, early detection, timely response, containment and elimination. The main goal of the strategy is to establish a robust global animal health system that effectively manages major animal health risks, paying particular attention to the animal–human-ecosystems interface and placing disease dynamics into the broader context of agriculture, socio-economic development and environmental sustainability, while recognising the essential role of adequate nutrition for health.

The Action Plan proposes specific actions in five technical work areas: (1) Understanding the cross-sectoral nature of health hazards; (2) Developing core technical capacities to deal with animal diseases at national, regional and global levels; (3) Strengthening health systems (institutional development and policy formulation); (4) Promoting animal health strategies that are socially acceptable and economically viable and (5) Fostering collaboration between animal, human and environmental health sectors. The Action Plan is upheld by three functional work areas: (A) Ensuring adequate human resources; (B) Communicating the Action Plan appropriately and (C) Establishing robust mechanisms for monitoring and evaluation. The Action Plan proposes to assist least developed countries to build their capacities in early warning, early detection and rapid response to disease outbreaks. The actions recommended are risk-based and tailored to the local context, engaging the people involved through participatory processes. All actions of the plan aim at sustainability and ownership by countries and regions and range from immediate to long-term actions with a developmental perspective.

The Action Plan is coherent with FAO's current strategic framework and several of its initiatives, including the strengthening of the Emergency Prevention System (EMPRES) for transboundary animal diseases, established in 1994 to combat threats to production, health and the environment. Since its endorsement, the FAO Programme Committee has requested two Action Plan updates to date. A planned submission for consideration to the Commission on Agriculture will be made by May 2014 to gain further support from FAO's governing bodies.

Operationalising One Health. FAO is involved in hundreds of field programmes around the world promoting and guiding food production, nutrition, policy, supporting sustainable forestry development and environmental issues, mitigating and adapting to the effects of climate change and responding to emergencies and crises. In its vast field programme, capacity development and outreach are key components. The farmer field schools and livestock field schools that incorporate the latest knowledge in crop production and animal health are one example, and with their constantly developing curricula, *One Health* could easily be incorporated by including hygiene and health as an integral part of farming, with inputs from

physicians with a thorough understanding of participatory approaches and community communication (Centres for Disease Control and Prevention 2012; Food and Agriculture Organisation 2011a). It is recognised that this effort is community-based and needs a long-term horizon, whereby the learning activities are undertaken and the positive experiences become second nature to those that are engaged to impart their
know-how to the next generation. One Health Clubs are another example, also at the village or community level, where human health is addressed as an expression of wider animal health and environmental health. One Health Clubs might focus on rabies and raise awareness about responsible dog ownership. These Farmer Field Schools and One Health clubs require monitoring, beginning with baseline information or participatory rural appraisals, where impacts can be evaluated and the platform improved, adapted and promoted for use elsewhere. Within professional competencies, a field/on-the-job cross-disciplinary training would be beneficial to further knowledge and to establish inter-sectoral links. At the professional level, modules and curriculum of the Field Epidemiology Training Programme (for physicians) can and have been adapted for veterinary laboratory specialists in Asia and Africa and the newer Field Epidemiology Training Programme for veterinarians (including components for wildlife specialists in other agencies outside health and agriculture), serve as proven models in increasing capacities and competencies while maintaining employment within the public sector (Minjauw et al. 2002; Food and Agriculture Organization 2009). In this continuing education programme, trainees participate in a 2-year programme spending some 25 % of their time in a classroom setting or with individualised or group mentor sessions, honing their skills on a pertinent problem within their community or country while still being active in the workplace. The programme counts on a cohort system, whereby those in the second year assist in teaching those in the first year (under the supervision of an experienced instructor, facilitator and epidemiologist). In addition, professionals and decision makers can advance *One Health* through their participation in simulation exercises (i.e., a potential zoonotic disease outbreaks that requires collaboration between public health, veterinary services, private sector, tourism authorities, national security and could involve communications professionals in a mock crisis situation and a pretend press conference). Lastly, operationalising *One Health* can well begin early in the academic curriculum of future professionals ranging from communications, to forestry and the environment as well as the biomedical sciences.

2 Conclusion

FAO's animal health portfolio stresses the importance of the broad, multi-disciplinary and multi-sectoral collaborative approach adopted by its Action Plan in attending to the health risks at the animal–human-ecosystems interface and contributes to the FAO/OIE/WHO tripartite vision. It will be important that FAO act on the priority activities within the Action Plan, ensure periodic monitoring of performance towards effective animal disease risk management at international, regional and country level and most importantly at the local level in a people-centred approach that will reflect and contribute towards achieving all of the eight millennium goals. FAO's *One Health* approach would reach beyond its animal health service and with corporate-wide support can promote the practices local, national and global level to ensure global health security.

References

Ashley S, Beazely P, Holden S (1999) Livestock in poverty-focused development. Livestock in Development, Crewkerne

Boyd J, Banzhaf S (2006). What are ecosystem services? Discussion paper 06–02. Resources for the Future, Washington, 26 pp

Centres for Disease Control and Prevention (2010) http://www.cdc.gov/globalhealth/fetp/. Accessed 1 June 2012

Food and Agriculture Organization (2009) State of food and agriculture: livestock in the balance. FAO, Rome

Food and Agriculture Organization (2011a) Emergency centre for transboundary animal diseases—RAP newsletter. http://www.fao.org/docrep/014/al909e/al909e00.pdf. Accessed 1 June 2012

Food and Agriculture Organization (2011b) Sustainable animal health and contained animal-related human health risks–in support of the emerging One-Health agenda. 106th Programme committee item 3, 12 pp. http://www.fao.org/docrep/meeting/021/ma145e.pdf. Accessed 10 May 2012

Food and Agriculture Organization, Vétérinaires Sans Frontières Belgium (2009) Pastoralist field school—guidelines for facilitation. FAO, Rome

Garrett KA, Mundt CC (1999) Epidemiology in mixed host populations. Phytopathology 89:984–990

Jones KE, Patel NG, Levy MA, Storeygard A, Balk D, Gittleman JL, Daszak P (2008) Global trends in emerging infectious diseases. Nature 451:990–993

Keesing F, Belden LK, Daszak P, Dobson A, Harvell CD, Holt RD, Hudson P, Jolles A, Jones KE, Mitchell CE, Myers SS, Bogich T, Ostfeld RS (2010) Impacts of biodiversity on the emergence and transmission of infectious diseases. Nature 468:647–652

Minjauw B, Muriuki HG, Romney D (2002) Adaptation of the farmer field school methodology to improve adoption of livestock health and production interventions. In: Responding to the increasing global demand for animal products. Proceedings of the British Society of Animal Science International Conference, Merida, Mexico, 12–15 Nov 2002

Polis GA, Anderson WB, Holt RD (1997) Toward an integration of landscape and food web ecology: the dynamics of spatially subsidized food webs. Annu Rev Ecol Syst 28:289–316

United Nations (2012) Millennium development goals. http://www.un.org/millenniumgoals/bkgd.shtml. Accessed 10 March 2012

Development of a One Health National Capacity in Africa

The Southern African Centre for Infectious Disease Surveillance (SACIDS) One Health Virtual Centre Model

Mark Rweyemamu, Dominic Kambarage, Esron Karimuribo, Philemon Wambura, Mecky Matee, Jean-Marie Kayembe, Aaron Mweene, Luis Neves, Justin Masumu, Christopher Kasanga, Bernard Hang'ombe, Kim Kayunze, Gerald Misinzo, Martin Simuunza and Janusz T. Paweska

Abstract Among the many challenges to health, infectious diseases stand out for their ability to have a profound impact on humans and animals. The recent years have witnessed an increasing number of novel infectious diseases. The numerous examples of infections which originated from animals suggest that the zoonotic pool is an important and potentially rich source of emerging diseases. Since emergence and re-emergence of pathogens, and particularly zoonotic agents, occur at unpredictable rates in animal and human populations, infectious diseases will

M. Rweyemamu (✉) · D. Kambarage · E. Karimuribo · P. Wambura · C. Kasanga · G. Misinzo
SACIDS at Sokoine University of Agriculture, Morogoro, Tanzania
e-mail: mark.rweyemamu@btinternet.com

M. Matee · K. Kayunze
SACIDS at Muhimbili University for Health and Allied Sciences, Dar es Salaam, Tanzania

J.-M. Kayembe · J. Masumu
SACIDS at School of Public Health, University of Kinshasa, Kinshasa, Democratic Republic of Congo

A. Mweene · B. Hang'ombe · M. Simuunza
SACIDS at School of Veterinary Medicine, University of Zambia, Lusaka, Zambia

L. Neves
SACIDS at Faculty of Veterinary Medicine, Eduardo Mondlane University, Maputo, Mozambique

J. T. Paweska
SACIDS at National Institute for Communicable Diseases of the National Health Laboratory Services, Sandringham, Johannesburg, South Africa

Current Topics in Microbiology and Immunology (2013) 366: 73–91
DOI: 10.1007/82_2012_244

Published Online: 22 July 2012

constitute a significant challenge for the public health and animal health communities in the twenty-first century. The African continent suffers from one of the highest burdens of infectious diseases of humans and animals in the world but has the least capacity for their detection, identification and monitoring. Lessons learnt from recent zoonotic epidemics in Africa and elsewhere clearly indicate the need for coordinated research, interdisciplinary centres, response systems and infrastructures, integrated surveillance systems and workforce development strategies. More and stronger partnerships across national and international sectors (human health, animal health, environment) and disciplines (natural and social sciences) involving public, academic and private organisations and institutions will be required to meet the present and future challenges of infectious diseases. In order to strengthen the efficiency of early warning systems, monitoring trends and disease prediction and timely outbreak interventions for the benefit of the national and international community, it is essential that each nation improves its own capacity in disease recognition and laboratory competence. The SACIDS, a One Health African initiative linking southern African academic and research institutions in smart partnership with centres of science excellence in industrialised countries as well as international research centres, strives to strengthen Africa's capacity to detect, identify and monitor infectious diseases of humans and animals, to better manage health and socio–economic risks posed by them, and to improve research capacity in investigation the biologic, socio–economic, ecologic and anthropogenic factors responsible for emergence and re-emergence of infectious diseases.

Contents

1 Introduction

A cursory examination of such sources as HealthMap, EMPRES-i and ProMed, which record disease events as reported in either official or unofficial publications shows a tendency for the clustering of disease events in North America, Europe and South-east Asia, a phenomenon that Jones et al. (2008) have referred to as the global richness map of the geographic origins of epidemic/emerging infectious disease (EID) events. It does not always reflect either source or areas of endemic settings of diseases, especially those of epidemic or transboundary nature. Studies which have focused on the risk of spread of EID (Brownlie et al. 2006, 2005; King et al. 2006; Rweyemamu et al. 2006; Jones et al. 2008; Woolhouse et al. Woolhouse 2008; FAO/OIE/WHO 2008; World Bank 2010) point to Africa and Asia as likely to harbour the endemic settings of both conventional and emerging epidemic diseases, especially in the human–livestock–wildlife interface areas. This view has been reinforced during a number of recent inter-governmental meetings (e.g.in Beijing in 2006, Sharm El-Sheikh in 2008, Winnipeg in 2009 and Stone Mountain in 2011). It is also implicit in the definition of One Health by the European Commission as "'the improvement of health and well-being through (i) the prevention of risks and the mitigation of effects of crises that originate at the interface between humans, animals and their various environments,.." There is also general conclusion that most emerging human pathogens have originated from animals or animal products and the risk of disease spread has been exacerbated by such factors as globalisation, climate change and systems of governance which do not target tackling pathogens and diseases at geographic source, e.g.in the endemic settings of developing countries or in animals. Yet, there is increasing evidence for the cost effectiveness for such targeted investment (Rushton et al. 2012; Jonas 2012).

Another observation is that Africa, probably has the highest burden of infectious diseases, with 72 % of the disease burden attributable to poverty, interactions between socio–economic opportunities and the health of animals, people and ecosystems, compared to 27 % in the rest of the world (Smith et al. 1999; Lopez et al. 2006; Kock et al. 2010; Muyembe et al. 2012). Also, such factors as climate variability, with fluctuations of drought and floods, internal displacement due to natural disasters or civil instability exacerbate the incidence, spread and socio–economic impact of disease in Africa, trapping many communities in chronic poverty.

These considerations are what propelled academic and research institutions in the Southern African Development Community (SADC) to set up the Southern African Centre for Infectious Disease Surveillance (SACIDS).

2 Brief History of SACIDS

The Foresight study on human, animal and plant infectious diseases identified future (to 2030) drivers of infectious disease risks (economic and climate change), future disease threats and future science/technologies that will help to manage the risk of infectious diseases within the evolving societal context of culture, governance, economic trends and human practices http://www.bis.gov.uk/foresight/our-work/projects/published-projects/infectious-diseases. Among its conclusions the following were the most pertinent to Africa:

- Many existing diseases will remain important, but new diseases will emerge in the future—noting that in the previous 25–30 years some 75 % of new/emerging infectious diseases of humans had originated from animals;
- Major infectious diseases are endemic in Africa and Asia; They constitute a high risk for future marginalisation of Africa;
- Human mobility and access to international markets for African animal and plant commodities could be severely constrained by infectious diseases in Africa; and
- Substantial advances in infectious disease prevention and management will be made through integration of research across sectors (human, animal, plant) and disciplines (natural and social science);
- New technological systems for early detection, identification and monitoring of infectious diseases have the potential to transform our capabilities in managing future disease risks, especially if challenges of international development are met. This convergence of technologies for DIM offers opportunity for innovative approaches in managing infectious disease risks in Africa.
- Societal contexts will be crucial in realising the benefits of the new technological systems.

In the light of these conclusions, a series of meetings in Africa, including the Foresight Africa workshop in Entebbe (Brownlie et al. 2005), the AU meeting of Directors for Livestock Development in Kigali 2004, the Congress of African Scientists and Policy Makers (CASP) in Alexandria, Egypt, 2006, the AU-Foresight meeting in Pretoria, September 2007 and the FAO-SADC-OIE meeting of SADC Chief Veterinary Officers (CVOs) plus regional experts in infectious diseases of livestock, wildlife and humans (zoonoses) in Arusha, Tanzania, August, 2007, all pointed to the need for urgent action for inter-sectoral initiatives aimed at accelerating the capacity of African institutions for the detection, identification and monitoring of infectious diseases. A common thread was that such initiatives should be rooted in national systems for disease surveillance and should foster inter-sectoral collaboration between the public health and animal health sectors.

These meetings and consultations culminated in a workshop in Pretoria from 22 to 25 January 2008, which outlined the Vision for a **Southern African Centre for Infectious Disease Surveillance (SACIDS)** as a virtual centre of academic and research institutions. The same workshop also recommended that in each participating country there should be a formal mechanism for promoting inter-sectoral collaboration on infectious diseases, as a national virtual centre for infectious diseases or **National Centre for Infectious Disease Surveillance (NatCIDS)**.

The January 2008 workshop of the founding SACIDS consortium members (i.e. academic and research medical and veterinary institutions of the Democratic Republic of Congo, Mozambique, Tanzania and South Africa) together with

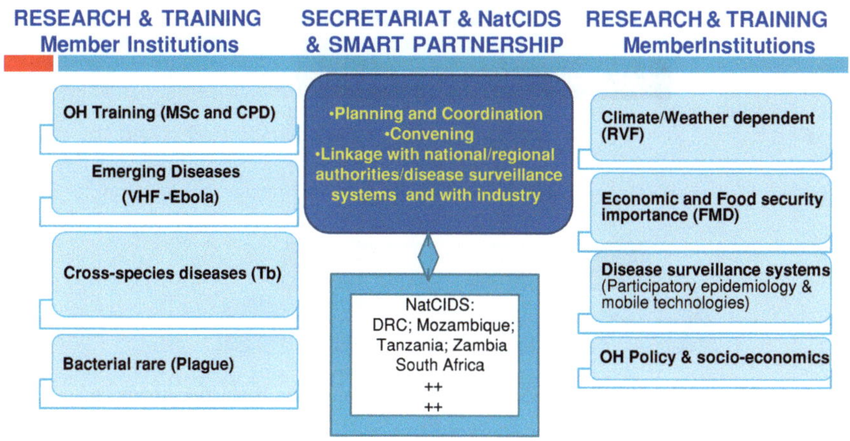

Fig. 1 The programmatic structure of SACIDS visual centre

potential Northern Partners and potential technical and financial enablers, as well as two African plant disease specialists, defined the Vision, Mission and key Objectives of SACIDS.

3 The SACIDS One Health Model

SACIDS is a One Health Virtual Centre linking southern African academic and research institutions involved with infectious diseases of humans, animals and ecosystems in smart partnership with centres of science excellence in industrialised countries as well as international research centres. The founding African institutions are from Tanzania, Democratic Republic of Congo (DRC), Mozambique, South Africa and Zambia and the Founding Smart Partnership institutions are the University of London colleges which are linked by the London International Development Centre, including the London School of Hygiene and Tropical Medicine, the Royal Veterinary College and School of Oriental and African Studies as well as the International Livestock Centre (ILRI) located in Kenya.

The mission of the centre is: to improve Africa's capacity to detect, identify and monitor (DIM) infectious diseases of humans, animals, ecosystems and their interactions in order to better manage the risks posed by them.

It seeks to implement this mission through its One Health focus, which SACIDS defines as addressing **infectious diseases** in the endemic settings of Sub-Saharan Africa through a collaborative effort between natural and social sciences to advance the understanding of interactions between humans, animals and the environment to improve public and animal health (Rweyemamu et al. 2012). This focus is in line with the definitions or hypothesis of advancing the One World, One

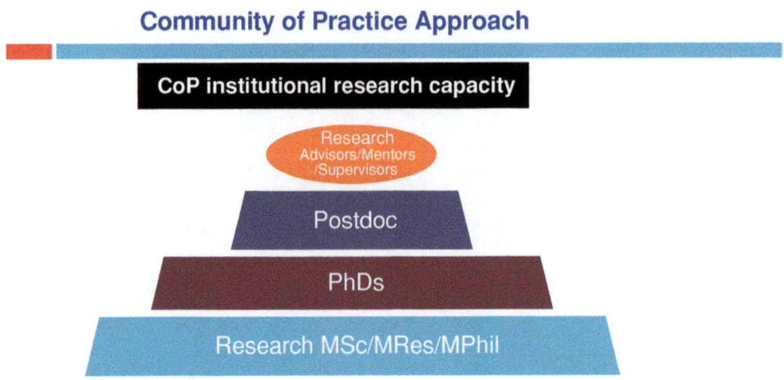

Fig. 2 RESEARCH APPRENTICESHIP through a community of practice approach

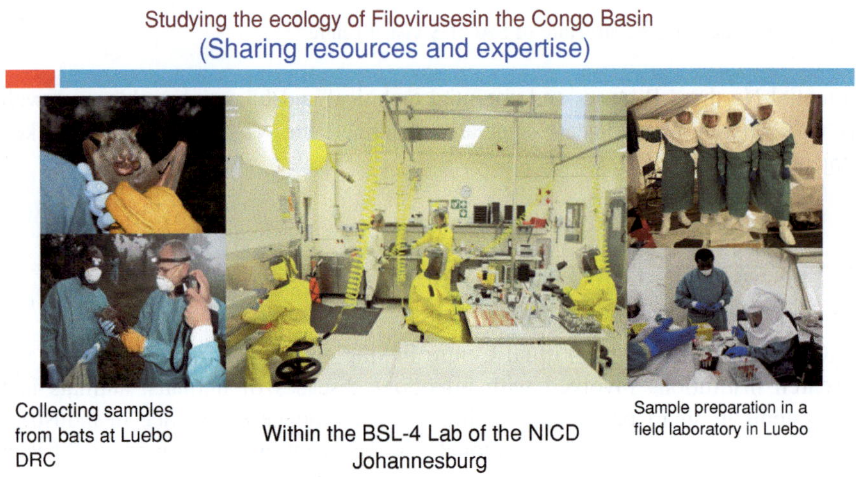

Fig. 3 Filovirus community of practice

Health approach (King 2008; FAO/OIE/WHO 2008; European Commission 2010; Parkes et al. 2005; Zinsstag et al. 2011).

The SACIDS is driven primarily by the objectives of institutional capacity development. It seeks to do so through a 3-pronged hypothesis namely that:

- The **Virtual Centre** concept would be more cost effective than a single physical centre for DIM of IDs
- **Theme Driven Programmes** would have a faster impact on capacity than isolated individual projects
- **The Community of Practice** approach would enhance the quality of supervision/mentoring of trainees as well as for research collaboration across institutions, projects and themes

3.1 The Basic Structure of SACIDS Virtual Centre

The overriding organisational structural principle is one of equity between the public health and animal (including wildlife) health sectors. This is also reflected in the governance structure of SACIDS. Thus the executive director is supported by two non-executive deputy directors, one who has an oversight over activities in the human health sector while the other focuses on the animal health sector.

In each of the four founding participating countries a National virtual Centre for Infectious Disease Surveillance (NatCIDS) exists, which links in-country institutions and the ministerial sectors responsible for human health, animal health and environmental health. In each country, where the coordinator is from the human health sector, the deputy is chosen from the animal health sector and vice versa. The programmatic structure of SACIDS is shown in Fig. 1. The central coordinating unit is the SACIDS Secretariat at the Sokoine University of Agriculture together with the NatCIDS in the participating countries. Smart Partners and the member institutions are involved in the execution of SACIDS programmes (Figs. 2, 3).

3.2 The Role of NatCIDS

The National Centres for Disease Surveillance (NatCIDS) play a key national role in guiding, leading, training and generation of scientific innovations for use in the surveillance and risk management strategies including early warning systems, emergency preparedness and interventions needed for prompt containment or control of infectious diseases of humans and animals. NatCIDS are made up of relevant experts, epidemiologists and stakeholders across the country and have a principal role in promoting multi-institutional, multi-sectorial partnership and collaboration both at national and international level in order to implement common activities and achieve strategic objectives of SACIDS.

In brief, the main activities of NatCIDS could be summarised as follows:

- To create a national platform to share and harmonise efforts from diverse sources with regard to infectious diseases surveillance and risk management
- To assist country diseases surveillance programme by evaluating and strengthening surveillance systems both on medical and veterinary sides with specific reference to the "One Health Concept"
- To work with stakeholders in preparing early, coordinated and accurate responses in case of epidemics or emerging disease occurrences
- To provide a better understanding of the circumstances of infectious diseases occurrence through state-of-the art involvement and investigations during outbreaks
- To source for resources through grants application for long self-sufficiency of the consortium at the national level
- To create effective collaboration with other African countries and Northern partners in addressing One Health issues

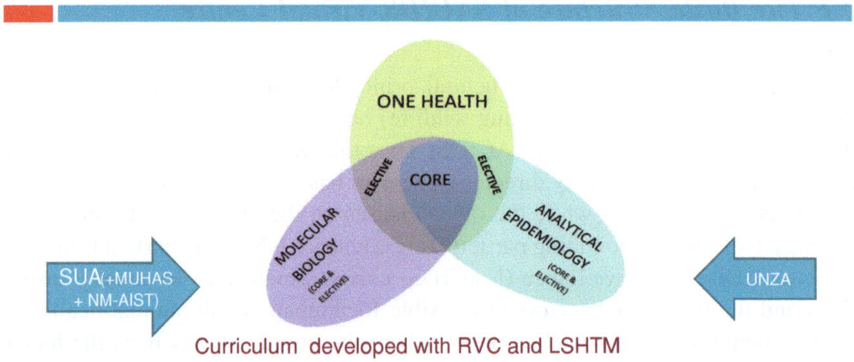

Fig. 4 Structure of SACIDS sponsored one health Msc At SUA and UNZA for Yr1 courses;(Yr2 devoted to research project)

3.3 Training for One Health

No country, rich or poor, is immune from the risk of microbial threats. Although specific conditions leading to emerging or re-emerging of infectious agents may be different, the challenge of managing disease spread is virtually the same, but not all countries have the same capacity to respond to this challenge. Within the "global village", individual countries/regions have different priorities, often contrastingly varying financial and infrastructure resources. However, an unprecedented amount of rapid movement of humans, animals and animal products between these countries/regions presents a biorisk not only to an individual country/region, but to the entire international community. The growing realisation that pathogens do not respect traditional epistemological divides has resulted in the emergency of the 'One Health' initiative to advocate for closer collaboration across the health disciplines and has provided a new agenda for health education. Not only has the need for interdisciplinary participation been acknowledged in projects involving control of zoonoses (Marcotty et al. 2009; Roth et al. 2003; Zinsstag et al. 2007) but there has been a new awakening to change the way health professionals are educated (Marcotty et al. 2009; Zinsstag et al. 2011). It is against this background that SACIDS included in its capacity development programme the development of two One Health driven 2-year MSc programmes, one at the School of Veterinary Medicine, University of Zambia, specialising in analytical epidemiology and the second at Sokoine University of Agriculture in Tanzania, specialising in molecular biology (Fig 4).

3.4 Developing One Health Research Capacity

The Community of Practice (Wenger et al. 2002; Rweyemamu et al. 2010) approach for stimulating research within the SACIDS consortium is used to develop capacity in seven research themes, each with a specific exemplar. These have been (i) Emerging viral disease diseases—represented by **Ebola and Marburg**; (ii) Climate dependent, vector-borne diseases (**Rift Valley fever**); (iii) Diseases with potential for inter-species concern/spread between wildlife, livestock and humans—represented by Tuberculosis; (iv) Diseases of economic and food security importance—represented by **foot-and-mouth disease**; (v) Bacterial rare diseases—represented by plague; (vi) Systems for disease surveillance and preparedness analysis— focusing on participatory epidemiology and the application of **mobile technologies** to field data capture and transmission; and (vii) socio–economic approaches to **One Health policy research**.

The geographical focus of our studies has been in the ecosystems with a high human–livestock–wildlife interaction or in cross-border areas. The Community of Practice for each theme has comprised a career development postdoctoral fellow (supported for 3–5 years), two or three PhD students supported for 4 years and a similar number of M.Phil. or Research MSc students supported for 2 years, all co-supervised and mentored by the same pool of specialists from both Southern Africa and the UK or ILRI (Fig. 2). Each CoP has been encouraged to seek collaboration with other groups working in the same countries on related objectives.

4 Case Studies

4.1 Discovery and Characterisation of Novel Arenaviruses *in Africa*

Emerging diseases, notably zoonoses caused by negative-stranded RNA viruses continue to be a formidable problem for public health and veterinary communities globally. Arenaviruses, principally rodent-borne viruses, are capable of causing severe syndrome of viral haemorrhagic fevers (VHFs) in humans. The family *Arenaviridae* includes 23 recognised species, of which 6 can cause outbreaks of VHFs with high case fatality rates. Until recently, Lassa virus was the only known arenavirus to cause VHFs in humans in West Africa (Buchmeier et al. 2007; Charrel et al. 2008). However, the importation of a previously unrecognised arenavirus to South Africa after air medical transfer of a critically ill patient from Lusaka, Zambia in September 2008 resulted in a dramatic VHF nosocomial outbreak in Johannesburg with a case fatality rate of 80 %. International collaboration during this outbreak allowed for rapid identification of the novel virus, thus reassuring that health and scientific communities are committed and have powerful tools to rapidly detect and respond to the challenges of emerging unknown pathogens. The history of the outbreak, however, is a serious warning that highly pathogenic arenaviruses could be more widely prevalent in Africa (Paweska et al. 2009). It also evidences that international movement of patients contributes to unintentional spread of dangerous pathogens with dramatic public health consequences. Genome molecular study, including the application of unbiased high-throughput sequencing, allowed for rapid characterisation of a new member of the family *Arenaviridae* (Briese et al. 2009) provisionally named Lujo virus (LUJV), but our knowledge on its ecology, epidemiology, including host range, natural transmission cycle and distribution is in paucity. In the light of an unusual highly mortality rate in patients infected with LUJV, studies on tissue tropism, dynamics of viral replication and dissemination, mechanisms of pathogenesis in rodent host species and other potential vertebrate hosts, including primates, are needed.

The natural reservoir host of LUJV remains unknown likewise the source and route of infection of the first (index) case. She kept domestic pets and horses on her agricultural holding near Lusaka and there was evidence of rodent activity in the stables. Rodents could contaminate objects left unattended on the ground by excretion of virus in urine, including broken glass which apparently resulted in a deep cut to her foot about 10 days before medical evacuation to South Africa (Paweska et al. 2009). These circumstances might be of potential epidemiological relevance as she developed first symptoms within the incubation period of arenavirus infection after the cut. The natural reservoirs of arenavirus in Africa are rodents of the family *Muridae*, especially *Mastomys natalensis*. To-date only non-pathogenic arenaviruses have been found in areas surrounding Zambia. A study on the prevalence of arenaviruses among *M. natalensis* rodents in Zambia was conducted less than 1 year after the LUJV outbreak, from May to August 2009, including areas surrounding the cities of Lusaka, Namwala and Mfuwe, but not specifically on the farm of the index case. Nevertheless, of the total of 263 rodents captured, 5 were positive for an arenavirus infection in kidneys, of which 17 % of the 23 rodents captured near Lusaka and 4 % of the 24 captured in Namwala were positive, but none of the 143 rodents captured in Mfuwe were positive for arenavirus. Phylogenetic analysis of the four Zambian arenavirus isolates showed distinct sequences between Old World and New World arenaviruses. The novel Lusaka and Namwala strains, collectively designated Luna virus (LUNV), are genetically different from the LUJV, but closely related to nonpathogenic arenaviruses that have been found from central to eastern Africa (Ishii et al. 2011).

4.2 Re-emergence of Rift Valley Fever in South Africa 2008–2011

Rift Valley fever (RVF) was first reported in South Africa when a large outbreak occurred during 1950 and 1951. Subsequently, outbreaks with confirmed human cases occurred in South Africa in 1953, between 1974 and 1976, and in 1999. Limited and isolated outbreaks occurred in South Africa in 2008 and 2009

(Swanepoel and Paweska 2011; Bret et al. 2011), followed by a large, widespread outbreak during 2010 and 2011. Between 2008 and 2011, more than 2000 specimens from suspected human RVF cases were submitted to the Special Pathogens Unit of the National Institute for Communicable Diseases of the National Health Laboratory Service for laboratory confirmation. In 2008 and 2009, a total of 24 non-fatal human RVF cases were laboratory confirmed. In 2010, a total of 241 human cases were confirmed of which 25 were fatal. No human fatalities were recorded among 37 confirmed cases during 2011. The outbreaks were geographically linked with outbreaks in domestic ruminants and occurred mostly on the inland plateau of the country, notably in the Free State, Northern Cape, North-West, Eastern Cape and Western Cape provinces. The RVF outbreaks in humans peaked in March 2010 when more than 100 cases were laboratory confirmed. No new cases have been recorded in 2012 to date. In total, 302 human cases were confirmed from 2008 to 2011. In the same period, a total of 13,902 animal cases were confirmed of which 8,581 were fatal. Animal cases involved primarily domestic ruminants, but wild animals, including buffalo, sable, nyala, alpaca, llama and Asian buffalo were also affected. Overlap of geographic distribution of animal and human cases indicates that most humans were infected through direct contact with infected animals. Indeed, 254 (89 %) of the confirmed cases reported a history of contact with animal tissues or bodily fluids. All fatal cases occurred in 2010 (case fatality ratio 10 %). The overall case fatality ratio from 2008 to 2011 was 8 % (Jansen van Vuren et al. 2012). The 2008 and 2009 outbreaks were caused by the virus genetic variants representing lineage C comprising virus isolates from Zimbabwe (1978–1979), Madagascar (1991), Kenya (1997–1998, 2006–2007), South Africa (1999) and Madagascar (2003). The 2010 and 2011 outbreaks were caused by virus genetic variants related to a Namibian isolate from of lineage H first identified in 2004. One isolate from the 2010 outbreaks was genetically distinct, and was closely related to the Smithburn neurotropic vaccine strain. The virus was isolated from a veterinarian who experienced a needle stick injury while vaccinating livestock that likely was already infected by wild-type circulating virus. This isolate is therefore likely a reassortant of the wild-type and the vaccine RVFV strains (Grobbelaar et al. 2011).

4.3 Studying the Ecology of Filoviruses in the Congo Basin

The sporadic outbreaks of filovirus infections in humans are believed to result from contact with an infected animal and subsequent transmission between persons by direct contact with infected blood or body fluids. Infected individuals succumbing to filovirus infection exhibit virus-mediated impairment of early innate immune responses allowing for rapid progression of filovirus infection. The unavailability of antiviral therapy or approved vaccines, and the elusive nature of the spillover of filoviruses from a reservoir source to humans hamper countermeasures to effectively prevent the severe course of filovirus disease and transmission. Viruses belonging to the filovirus group were first discovered in 1967 for Marburg virus (Saijo et al. 2006) and in 1976 for Ebola virus (WHO 1978a, b). For a long time, the epidemiological circumstances surrounding filovirus outbreaks suggested that bats may have served as the primary source of infection in humans and non-human primates. Despite intensive efforts to trap thousands of vertebrate and invertebrate hosts in filovirus outbreak areas, isolation of live Ebola virus was unsuccessful, and the ecology and epidemiology of filoviruses are still not well understood. This is mostly because filovirus outbreaks occur irregularly, in poor resource, and remote areas of Africa, consequently investigations of filovirus outbreaks are hugely delayed and dependent on international support. In addition, handling of filoviruses requires the use of biosafety level four (BSL 4) facilities which are unavailable in these countries. Therefore, investigation of filovirus outbreaks, biology, ecology and epidemiology of filoviruses require a collaborative approach involving local researchers and overseas partners in order to secure required funding, training, diagnostic materials and laboratory support.

The SACIDS study of the ecology of filoviruses is undertaken in the Congo basin and particularly in the Democratic Republic of Congo (DRC) where several filovirus outbreaks occurred in the past. The programme is based on a collaborative approach involving several Congolese institutions (e.g. National Institute of Biomedical Research, Veterinary Laboratory of Kinshasa, University of Kinshasa), the National Institute of Communicable Diseases (NICD) a branch of the National Health Laboratory Services of South Africa plus the World Health Organisation (WHO). In this collaborative approach, the Congolese institutions provide administrative and human resources for sample collection in the field, the SACIDS, the NICD/NHLS and the WHO provide maximum security laboratory facility, training, financial and logistic support. Specific objectives of these studies are to collect samples from putative filovirus reservoirs in the DRC, notably from targeted bat species, and conduct their laboratory analysis in the BSL4 facility at NICD/NHLS (Fig. 2). These studies are based on the hypothesis of bats being reservoirs of Ebola virus and thus aim at isolating Ebola virus from bat organs.

As part of this programme, several other studies are conducted by the SACIDS postdoctoral fellow, PhD (2) and MSc (2) students to improve the understanding of the epidemiology of Ebola disease, develop rapid test for disease detection and a bat cell culture technique for virus isolation and assess the disease pathogenesis in the host. Results of these studies will assist in improving the prevention and further response strategies to control these diseases.

4.4 Investigation of the Presence of the Plague Bacillus in Rodents and Fleas in Zambia and Tanzania

Rodents and their flea ectoparasites have been known to harbour viral and bacterial pathogens from discrete ecological foci worldwide. The close proximity of humans to rodents and fleas are a recognised route of transmission of pathogens. The increase in human population density has led to people moving in new areas where they are at risk of rodent and flea-borne pathogens. A classic example of such pathogens is the plague bacillus *Yersinia pestis* which is usually transmitted by fleas. The disease is enzootic in many parts of Asia, Africa and America (Stenseth et al. 2008; WHO 2000). Major outbreaks are regularly documented in Africa where living conditions support close association of rodents, fleas and humans. Of significance are rodent species outside the peri-domestic environment which may act as important liaison hosts between the sylvatic reservoir and humans. To enhance understanding of the route and transmission of the plague bacillus in such environments, field studies are being undertaken in Tanzania and Zambia. The studies are focused on analysing the presence of *Yersinia pestis* in both rodents and fleas (Hang'ombe et al. 2012) by using molecular methods which are easier and safer to handle, as *Yersinia pestis* requires appropriate bio-containment facilities (Hinnebusch et al. 1998). Areas known to be endemic with plague are being targeted for surveillance. In these areas, rodents, fleas and where possible human patients are targeted for *Yersinia pestis* detection. In addition, the best possible options for sample collection are being designed for the benefit of the communities living in these remote and inaccessible areas. The results obtained from such studies will have a direct impact on disease control strategies.

Based on the same protocol template within the same ecosystem, other pathogens are also being investigated simultaneously with *Yersinia pestis*. So far, studies have revealed the presence of a novel arenavirus from the rodents (Ishii et al.2011). Furthermore, the protocols that have been established for *Yersinia pestis* have also been successfully applied in the control of *Bacillus anthracis* outbreaks in Zambia.

4.5 Collaborative Study for Diseases that Threaten Food Security and Livelihood

Foot-and-mouth disease (FMD) was selected as the exemplar of our One Health approach which focuses on those diseases which threaten food security and the livelihoods communities (Perry et al. 2008). FMD is a highly contagious and economically devastating disease of cloven-hooved domestic and wild animals. FMD causes mortality among young animals and contributes to production losses in adults thereby threatening food security and livelihood. The causative agent of FMD is FMD virus (FMDV), which belongs to the genus *Aphthovirus* in the family *Picornaviridae*. There are seven immunologically distinct FMDV serotypes (O, A, C, Southern African Territories (SAT 1, SAT 2, SAT 3 and Asia 1) known to exhibit different global distributions with defined epidemiological clusters (Rweyemamu et al. 2008) and topotypes within each serotype (Knowles et al. 2003; Vosloo et al. 2002). In Southern Africa, FMD is recognised as a disease of strategic importance for which there is regional agreement and commitment for its progressive control. Accordingly, the region has developed in conjunction with the FAO, OIE and AU-IBAR, the SADC FMD Progressive Control Pathway (SADC Report, 2011). Southern Africa relies profoundly on livestock production as a source of economic growth and livelihoods for the rural poor with nearly over 50 % of the Southern African population deriving their livelihoods from livestock (Perry and Rich 2007). Despite its importance in Southern Africa, the epidemiology of FMDV in the SADC region has not been deeply and consistently studied. A study to investigate the complex epidemiology and molecular determinants of FMD endemicity in Southern Africa is being undertaken through SACIDS with the key research question "*what contributes to FMD endemicity in Southern Africa and what options exist for its risk management*"? This study is carried out by employing the community of practice approach, in collaboration with other initiatives such as SADC-TADs, BBSRC-CIDLID and DANIDA funded FMD projects in East Africa. So far, four serotypes (A, O, SAT 1 and SAT 2) and respective topotypes have been identified in different geographic areas in Tanzania using antigen ELISA, NSP ELISA, RT-PCR and sequencing of FMDV capsid protein genes (Kasanga et al. 2012: Manuscript in submission). The findings of this research will unravel the geographical distribution, genetic diversity and antigenicity of circulating FMDV strains, epidemiological and molecular determinants of FMD endemicity, and evolutionary characteristics of FMDV field strains. This information is useful in designing and recommending the appropriate strategies to implement for prevention and control of FMD in Africa with an ultimate increase in animal production and hence food security and improved livelihood. Similar approaches are to be employed in the future for studying other diseases that threaten food security and livelihood in the region such as Peste des Petits Ruminants (PPR) and African Swine Fever (ASF).

4.6 Developing Disease Surveillance Systems in the Human-Livestock-Wildlife Interface and Cross-Border Areas

Using participatory approach, we designed a surveillance system for diseases in animal and human populations in three ecosystems namely Ngorongoro, Kagera river basin and Zambezi river basin. Joint efforts coordinated by Southern African Centre for Infectious Disease Surveillance (SACIDS), East African Integrated Disease Surveillance network (EAIDSNet) and National Centre for Infectious Disease Surveillance (NatCIDS) in Tanzania and Zambia resulted in the development of One Health disease surveillance strategy for the three ecosystems. The Ngorongoro ecosystem represents an area of maximum human–wildlife–domestic animal interactions. This ecosystem is predominantly inhabited by the Maasai pastoral communities who keep cattle, goats and sheep and are in close proximity with wild animals in the wildlife protected areas of Ngorongoro Conservation Area

(NCA). The other two sites (Kagera and Zambezi River basins) represent cross-border ecosystems, where OH surveillance could be potentially effective in diagnosing and managing infectious diseases across borders. The Kagera river basin ecosystem is located in the Great Lake Region of eastern Africa and is shared between Uganda, Rwanda, Burundi and Tanzania. The Zambezi River basin is located in southern Africa and is shared between Zambia, Angola, Namibia, Botswana and Zimbabwe. The One Health disease surveillance strategy should consist of two complementing systems namely: (a) Community-based Active Surveillance (CAS) system designed to actively capture disease events in animal and human populations using simple case definitions of clinical signs and syndromes occurring in communities. This system uses community-based health reporters (CHRs) who would actively screen for the occurrence of disease events in human, wildlife and domestic animal populations. Data on these events are recorded and transmitted through Android mobile phones using the Epicollect data capture application in near to real time, and: (b) District-based Passive Surveillance (DPS) system uses existing surveillance strategies in animal and human (IDSR) health sectors with enhanced performance through application of mobile technologies in transmission of near to real-time data in the two health sectors.

Collaborating with other institutions in the United Kingdom (Royal Veterinary College and Imperial College London) as well as those in South-east Asia (BIOPHICS, Ministry of Public Health Thailand, MBDS and InSTEDD, Cambodia) assisted in the improvement of the OH surveillance strategy developed by SACIDS. The two systems (CAS and DPS) are linked together at the data analysis point. Data collected through CAS and DPS systems from remote sites are stored centrally on a server located at the SACIDS headquarters (for the Ngorongoro ecosystem) while those from the Zambezi river ecosystem are stored on a central server located at the University of Zambia (UNZA). The data storage points act as custodians and store data on behalf of the ministries responsible for human and animal health in respective countries. Data are analysed and summarised as reports that are shared with the two ministries and field-based disease management units at district headquarters.

4.7 Towards One Health Policy Research

The concept of One Health has recently received immense regional and international attention and this has led to the formation of one health initiatives to spearhead the agenda and promote inter-sectoral collaboration. Mainstreaming of One Health approaches in human and animal health programmes is expected to result in the development of integrative and effective risk management strategies for enhanced attainment of optimal health for the people, animals, plants and the environment. In addressing these goals, the SACIDS has as one of its seven research themes a Community of Practice (CoP) project on socio–economic approaches, including health policy research. The sub-themes of the CoP are (i) health policies, programmes/strategies and governance structures, and how they influence human and animal health service delivery systems; (ii) the scope, nature and factors influencing inter-sectoral collaboration, especially for infectious zoonotic disease surveillance and control; (iii) socio–economic impacts of infectious diseases and their control; and (iv) influence of health care seeking behaviour, health policies and service delivery systems on human and animal health. It is envisaged that the thrust of this CoP will generate additional information needed to facilitate the creation of a sound platform for enhanced inter-sectoral collaboration in infectious disease detection, identification, monitoring and control.

5 Towards one Africa, One Health

As a Virtual Centre, the SACIDS set out, initially, to focus on just the SADC region of Africa in the expectation that similar regional Virtual Centres will emerge in East Africa (EACIDS), West Africa (WACIDS), etc., all under the umbrella of the African Union. During the 2005 African infectious disease meeting in Entebbe, specialists called for **Pan-African Vision for Infectious Disease Management** (Brownlie et al. 2006) as:

"A Pan-African concerted effort, shared by AU member governments, reflecting the needs of African society and supported by the international community, with the goal of a society protected from the ravages of dangerous infectious diseases that compromise either human health or livelihoods and agriculture and economic development."

This was further elaborated by Rweyemamu et al. (2006) in assessment of future risks of infectious diseases of humans, animals and plants in Africa. The concept was further discussed at the 3rd Ordinary Session of the African Ministerial Conference on Science Technology in Mombasa, Kenya, November 2007. Since 2008, the AU has elaborated a policy framework which embraces this vision. While EACIDS and WACIDS are still at an embryonic stage, there have been several One Health-based developments in Africa either as funded projects, such as the Wellcome funded Afrique One project or organisational or institutional re-alignment as the USAID funded and stimulated organisation of the Deans of Veterinary and Public Health Faculties in Central and East Africa to promote one health training (OHCEA), or the new one health emphasis of the African Field Epidemiology Network (AFENET) or the East African Integrated Disease Surveillance Network (EAIDSNet) of the East African Community. Within the African Union structures, the AU Inter-African Bureau for Animal Resources (AU-IBAR) is developing a key thrust based on one health approaches, through lessons learnt from its avian influenza programme. It is also encouraging to note that one of the new programmes of the International Livestock Research Institute (ILRI), whose HQ is in Nairobi, Kenya relates to one health approaches. Several donor funded projects operating in Africa also have a One Health focus. Meanwhile, the mission of SACIDS has now expanded to make

SACIDS relevant to the whole of sub-Saharan Africa, albeit with its primary focus remaining southern and East Africa.

The goal of One Africa, One Health requires an African Forum that brings together different players on a common platform. To this end, SACIDS convened the first One Health Conference in Africa in July 2011 http://www.ojvr.org/index.php/ojvr/issue/view/33. But neither SACIDS nor any of the above-mentioned initiatives is, at the moment, adequately funded for such regular convening.

Acknowledgments We acknowledge the financial support from the Wellcome Trust, Rockefeller Foundation, Google Foundation and the UK Foresight programme. As a Virtual Centre, SACIDS heavily depends on the commitment and support of the host institution (Sokoine University of Agriculture) and all the member institutions in the DRC, Mozambique, South Africa, Tanzania and Zambia as well as the mentoring support and commitment of our Smart Partners, especially the London School of Hygiene and Tropical Medicine, the Royal Veterinary College and the London International Development Centre as well as the International Livestock Research Institute (ILRI).

References

Ishii A, Thomas Y, Moonga L, Nakamura I, Ohnuma A, Hang'ombe BM, Takada A, Mweene A, Sawa H (2011) Novel arena virus, Zambia. Emerging infectious diseases, 17:1921–1924

Brett NA, Weyer J, Paweska J, Nkosi D, Leman P, Tint KS, Blumberg L (2011) Outbreak of Rift Valley fever virus affecting veterinarians and farmers in South Africa, 2008, S Afr Med J 101:263–266

Briese T, Paweska JT, McMullan LK, Hutchison SK, Street C, Palacios G, Khristova ML, Weyer J, Swanepoel R, Egholm M, Nichol ST, Lipkin WI (2009) Genetic detection and characterization of Lujo virus, a new hemorrhagic fever-associated arenavirus from southern Africa. PLoS Pathog 5:e1000455

Buchmeier MJ, de la Torre J. C, Peters CJ (2007) Arenaviridae: the viruses and their replication. In: Knipe DM, Howley PM (eds) Fields Virology, 5th ed. Philadelphia: Lippencott Williams & Wilkins, Hagerstown pp1791–1827

Brownlie J, Peckham C, Waage J, Woolhouse M, Lyall C, Meagher L. Tait J, Baylis M, Nicoll A (2006) Foresight. Infectious diseases: preparing for the future. Future Threats. Office of Science and Innovation, London p 1–83. http://www.bis.gov.uk/assets/bispartners/foresight/docs/infectious-diseases/t1.pdf. Accessed January 2012

Brownlie J, Morgan D, Otim-Nape W, Rweyemamu M, Serwadda D, Waage J (2005) Infectious diseases in Africa: using science to fight the evolving threat report of a pan-African workshop in Entebbe, Uganda, August 2005. Office of Science and Innovation, London. http://www.bis.gov.uk/assets/bispartners/foresight/docs/infectious-diseases/a4.pdf (Accessed May 2012)

Charrel RN, de Lamballerie X, Emonet S (2008) Phylogeny of the genus Arenavirus. Curr Opin Microbiol 11:362–368

European Commission (2010) Outcome and impact assessment of the global response to the avian influenza crisis 2005–2010. Publications Office of the European Union, Luxembourg

FAO/OIE/WHO (2008) Contributing to one world, one health: a strategic framework for reducing the risks of infectious diseases at the animal-human-ecosystems Interface. OIE/WHO/UNICEF/World Bank and UN system influenza coordination, international ministerial conference on avian and pandemic influenza. Sharm El-Sheikh, October 14, 2008. http://un-influenza.org/files/OWOH_14Oct08.pdf Accessed January, 2012

Grobbelaar AA, Weyer J, Leman PA, Kemp A, Paweska JT, Swanepoel R (2011) Molecular epidemiology of Rift Valley fever virus. Emerg Infect Dis 17:2270–2276

Hang'ombe BM, Nakamura I, Samui KL, Kaile D, Mweene AS, Kilonzo BS, Sawa H, Sugimoto C, Wren BW (2012) Evidence of *Yersinia pestis* DNA from fleas in an endemic plague area of Zambia. BMC research notes. 5:72

Hinnebusch JB, Gage KL, Schwan TG (1998) Estimation of vector infectivity rates for plague by means of a standard curve-based competitive polymerase chain reaction method to quantify *Yersinia pestis* in fleas. 1998. Am J Trop Med Hyg, 58:562–569

Ishii A, Thomas Y, Moonga L, Nakamura I, Ohnuma A, Hang'omebe A, Takada A, Mweene B, Sawa H (2011) Novel arenavirus, Zambia. Emerg Infect Dis 17:1921–1924

Jansen van Vuren P, WeyerJ, Kemp A, le Roux C, Leman P, Grobbelaar A, Lekhuleni P, Dermaux-Msimang V, Archer B, Thomas J, Blumberg L, Paweska J (2012) Viral haemorrhagic fever outbreaks, South Africa, 2011. National Health Laboratory Service, National Institute for Communicable Diseases 1 April, 2012, Commun Dis Surveill Bull 10(1):1–5

Jonas O (2012) Economics of one health. Presentation to the one health summit 2012. Davos, 19–23 February, 2012

Jones KE, Patel N, Levy MA, Storeygard AD, Balk JL, Gittleman, Daszak P (2008) Global trends in emerging infectious diseases. Nature 451:990–993

Kasanga CJ, Sallu R, Kivaria F, Mkama M, Masambu J, Yongolo M et al (2012) Foot-and-mouth disease virus serotypes detected in Tanzania from 2003 to 2010: Conjectured status and future prospects', Onderstepoort J Vet Res 79(2), Art. #462, 4 pages. http://dx.doi.org/10.4102/ ojvr.v79i2.462

King DA, Peckham C, Waage JK, Brownlie J, Woolhouse M (2006) Infectious diseases—Preparing for the future, 8th Sept 2006. Science 303:1392–1393

King L (2008) The convergence of human and animal health one world–one health. UMN Workshop; Minneapolis, 14 May 2008

Knowles NJ, Samuel AR (2003) Molecular epidemiology of foot-and-mouth disease virus. Virus Res 91:65–80

Kock RA, Kock MD, Cleaveland S, Thomson G (2010) Health and disease in wild rangelands. In: Wild Rangelands Conserving wildlife while maintaining livestock in semi-arid ecosystems. Du Toit JT, Kock RA, Deutsch JC (eds). Wiley Blackwell Oxford pp 98–128

Lopez AD, Mathers CD, Ezzati M, Jamison DT, Murray CJL (2006) Global and Regional Burden of Disease and Risk Factors, 2001: systematic analysis of population health data. The Lancet 367:1747–1757

Marcotty T, Matthys F, Godfroid J, Rigouts L, Ameni G, van Pittius NG, Kazwala R, Muma J, van Helden P, Walravens K, de Klerk LM, Geoghegan C, Mbotha D, Otte M, Amenu K, bu Samra N, Botha C, Ekron M, Jenkins A, Jori F, Kriek N, McCrindle C, Michel A, Morar D, Roger F, Thys E, van den Bossche P (2009) Zoonotic tuberculosis and brucellosis in Africa: neglected zoonoses or minor public-health issues? The outcomes of amulti-disciplinary workshop. Ann Trop Med Parasitol 103:401–411

Muyembe-Tamfum JJ, Mulangu S, Masumu J, Kayembe JM, Paweska JT (2012) Ebola virus outbreaks in Africa: past and present. Onderstepoort J vet Res 79(2), Art. #451, 8 pages. http://dx.doi.org/10.4102/ojvr.v79i2.451

Parkes MW, Bienen L, Breilh J, Hsu L-N, McDonald M, Patz JA, Rosenthal JP, Sahani M, Sleigh A, Waltner-Toews D, Yassi A (2005) All hands on deck: transdisciplinary approaches to emerging infectious disease. EcoHealth 2:258–272

Paweska JT, Sewlall NH, Ksiazek TG, Blumberg LH, Hale MJ, Lipkin WI, Weyer J, Nichol PE, Rollin LK, McMullan CD, Paddock T, Briese J, Mnyaluza T -H., Dinh V, Mukonka P, Ching A, Duse G, Richards G, de Jong C, Cohen B, Ikalafeng C, Mugero C, Asomugha MM, Malotle DM, Nteo E, Misiani R, Swanepoel SR, Zaki, Members of the outbreak control, and investigation teams (2009) Nosocomial outbreak of novel arenavirus infection, Southern Africa. Emerg Infect Dis 15:1598–1602

Pepin M, Bouloy M, Bird B, Kemp A, Paweska J (2010) Rift Valley fever virus (*Bunyaviridae: Phlebovirus*): an update on pathogenesis, molecular epidemiology, vectors, diagnostics and prevention. Vet Res 41:61. doi:10.1051/vetres/2010033

Perry BD, Rich KM (2007) Poverty impacts of foot-and-mouth disease and the poverty reduction implications of its control, Veterinary Record 2007;160:238–241

Rushton J, Häsler B, de Haan N, Rushton R (2012). Economic benefits/drivers of a "one health" approach: why should anyone invest? Onderstepoort J vet Res 79(2), Art. #461, 5 pages. http://dx.doi.org/10.4102/ojvr.v79i2.461

Rweyemamu M, Otim-Nape W, Serwadda D (2006) Foresight. infectious diseases: preparing for the future, Africa Office of Science and Innovation, London p1–120. http://www.bis.gov.uk/assets/bispartners/foresight/docs/infectious-diseases/a1_id_africa.pdf. Accessed January 2012

Rweyemamu M, Roeder P, MacKay D, Sumption K, Brownlie J, Leforban Y (2008) Planning for the progressive control of foot-and-mouth disease worldwide. transboundary and emerging diseases.55:73–87

Rweyemamu MM, Paweska J, Kambarage D, Namuba F (2012) Towards One Africa, One Health: the SACIDS one health focus on infectious diseases. Onderstepoort J vet Res 79(2), Art. #449, 2 pages. http://dx.doi.org/10.4102/ojvr.v79i2.449

Stenseth NC, Atshabar BB, Begon M, Belmain SR, Bertherat E, Carniel E, Gage KL, Leirs H, Rahalison L (2008) Plague: past, present and future. PLoS Med 5:e3. doi:10.1371/journal.pmed.005003

Swanepoel R, Paweska JT (2011) Rift Valley fever. Oxford textbook of zoonoses: biology, clinical practise, and public health control 2nd edn. In: Palmer SR, Soulsby L, Torgerson PR, Brown DWG (eds), Oxford University Press, Oxford, pp 421–431

Saijo M, Niikura M, Ikegami T, Kurane I, Kurata T, Morikawa S (2006) Laboratory diagnostic systems for Ebola and Marburg hemorrhagic fevers developed with recombinant proteins. Clin Vaccine Immunol 13:444–51

Smith KR, Corvalan CF, Kjellstrom T (1999) How much ill health is attributable to environmental factors? Epidemiology 10:573–584

Vosloo W, Baston ADS, Sangare O, Hargreaves SK, Thomas GR (2002) Review of the status and control of foot and mouth disease in sub-saharan Africa. Rev. Sci. Tech. Off. Int. Epizoot. 21:437–449

Wenger E, McDermott R, Synder W (2002) Cultivating Communities of Practice: a guide to managing knowledge. Harvard Business School Press, Boston

WHO (1978a) Ebola haemorrhagic fever in Sudan. 1976. Bull World Health Organ 56:247–270

WHO (1978b) Ebola haemorrhagic fever in Zaire. 1976. Bull World Health Organ 56:271–293

WHO (2000) Human plague in 1998 and 1999. Weekly Epidemiol Rec 75:338–343

Woolhouse MEJ (2008) Epidemiology: Emerging diseases go global. Nature 451:898–899

World Bank (2010) People, pathogens, and our planet. Volume one: towards a one health approach for controlling zoonotic diseases. © World Bank

Zinsstag J, Schelling E, Roth F, Bonfoh B, de Savigny D (2007) And Tanner, M. Emerg Infect Dis 13:527–531

Zinsstag J, Schelling E, Waltner-Toews D, Tanner M (2011) From "one medicine" to "one health" and systemic approaches to health and well-being. Prev Vet Med 101:148–56

The Development of One Health Approaches in the Western Pacific

Ben Coghlan and David Hall

Abstract The Western Pacific Region, the most populous of six regional groupings of World Health Organization (WHO) member states, has seen the emergence of a series of novel zoonotic infections in the last decade. This has focused attention on addressing underlying risks and vulnerabilities in the complex interactions among people, animals, and environments as a better way to counter emerging diseases. This "One Health" approach is pertinent to the region because, it is a "hot spot" for the emergence of novel diseases from wildlife, because unexpected epidemics of re-emerging zoonotic diseases have caused morbidity and mortality in urban and periurban communities, and because it remains a sanctuary for well-known zoonotic infections. In this chapter, selected regional, multicountry, and national steps to operationalize One Health are discussed. While the region is well positioned to exploit the opportunities that have come with outbreaks of new diseases, the array of disconnected and overlapping initiatives from various consortia, donors, research institutes, and UN agencies is to some extent impeding the development of better ways of managing both new and old infections for the local, regional, and global good.

Contents

B. Coghlan (✉)
Burnet Institute, Melbourne, Australia
e-mail: coghlan@burnet.edu.au

D. Hall
University of Calgary, Calgary, Canada
e-mail: dchall@ucalgary.ca

Current Topics in Microbiology and Immunology (2013) 366: 93–111
DOI: 10.1007/82_2012_270

Published Online: 12 October 2012

1 Introduction

The Western Pacific Region is one of six regional groupings of World Health Organization (WHO) member states. It is the most populous region with over one-quarter of the global population living in 37 countries and territories (World Health Organization, Western Pacific Region 2012). These countries are diverse: from China with the world's largest population to some of the world's smallest states like the Pacific Islands of Niue, Tokelau, Nauru, and Tuvalu (Population Reference Bureau 2011); from one of the most densely populated countries, Singapore, to the least densely populated country Mongolia (Population Reference Bureau 2011); from highly developed states such as Australia, Japan, and the Republic of Korea to countries ranked among the world's least developed like Papua New Guinea and the Solomon Islands (UNDP 2011) (Table 1).

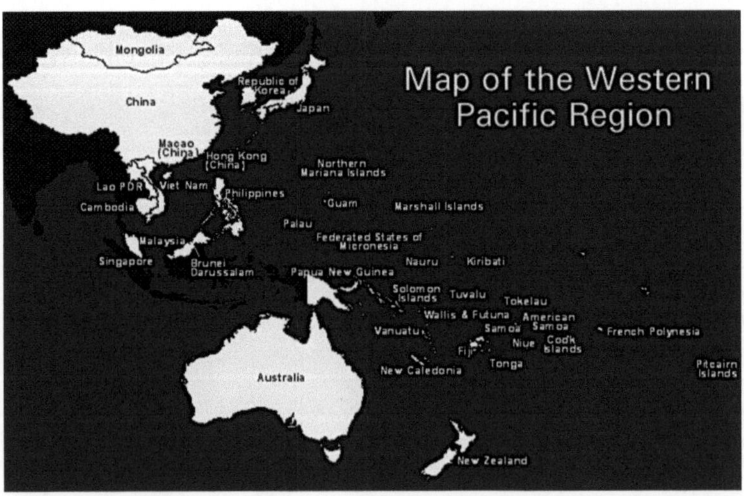

Table 1 Countries and Territories in the Western Pacific Region

1. American Samoa	11. Japan	21. New Zealand	29. Samoa
2. Australia	12. Kiribati	22. Niue	30. Singapore
3. Brunei Darussalam	13. Lao People's Democratic Republic	23. Commonwealth of the Northern Mariana Islands	31. Solomon Islands
4. Cambodia			32. Tokelau (New Zealand)
5. PR China	14. Macao (China)		
6. Cook Islands	15. Malaysia	24. Palau	33. Tonga
7. Fiji	16. Marshall Islands	25. Papua New Guinea	34. Tuvalu
8. French Polynesia (France)	17. Micronesia, Federated States of	26. Philippines	35. Vanuatu
		27. Pitcairn Islands (UK)	36. Vietnam
9. Guam (USA)	18. Mongolia	28. Republic of Korea	37. Wallis & Futuna
10. Hong Kong (China)	19. Nauru		
	20. New Caledonia (France)		

The emergence of a series of novel zoonotic infections from the region in the last decade triggered an unprecedented mobilization of the international public health community to address these threats. SARS in 2003, exposed weaknesses in national capacities to quickly identify, contain, and control a novel infection; these weaknesses equate to a persisting global threat. In 1997 and then again in 2004, bird flu (Influenza A/H5N1), the largest epizoonosis ever recorded, sounded a second call for global pandemic preparedness highlighting not only the shortcomings of human health services but the challenges of strengthening animal health and production systems, of restructuring food supply chains, and of sustaining responses for years. The virus remains endemic in poultry in China and Vietnam and has demanded far more than just emergency responses.

There is recognition that strategies to reduce the likelihood of disease emergence and transmission by addressing underlying risks and vulnerabilities in the complex interactions among people, animals, and environments, between human systems and natural ecosystems, may be a better way to counter emerging diseases. This has been referred to in various contexts as ecohealth (Charron 2012), particularly where it includes consideration of the role of environmental factors, and as a One Health approach in which health disciplines work together rather than in exclusion. This "One Health" approach is pertinent to the Western Pacific Region for three reasons.

First, the Mekong subregion within the Western Pacific Region has been designated a "hot spot" for the emergence of novel diseases from wildlife because of an amalgam of related anthropogenic drivers of disease emergence: rapid economic development, urbanization, advancing farming systems, demand for livestock products and deforestation, as well as population increases and aging (Jones et al. 2008). These factors cannot be addressed by the human or animal health sectors alone, necessitating a collective engagement with a range of sectors and communities.

Second, unexpected epidemics of re-emerging zoonotic diseases including rabies, anthrax, and leptospirosis have caused morbidity and mortality in urban and peri-urban communities in the Western Pacific Region. Some of these epidemics are being addressed using One Health approaches, and indicate the value in learning from and working with partners in the region when developing public awareness and preparedness plans for emerging infectious diseases (EIDs).

Third, the region remains a sanctuary for well-known zoonotic infections such as brucellosis that have been eliminated in many parts of the world but may be effectively addressed with an approach that is better tuned to tackle the complexities of real-world problems (World Bank 2009).

It is fitting then, that serious global commitment to this nascent approach was made in the region in Hanoi in 2010: the International Ministerial Conference on Animal and Pandemic Influenza aimed to "set the scene for a worldwide effort, over

Table 2 Selected global One Health initiatives and organizations implementing One Health activities in the Western Pacific Region

Chatham House Centre on Global Health Security	Global Initiative for Food Systems Leadership (GIFSL)
Centers for Disease Control and Prevention (CDC) One Health Office (OHO)	Health and Ecosystems: Analysis of Linkages (HEAL)
Connecting Organizations for Regional Disease Surveillance (CHORDS)	International Development Research Centre (IDRC)
Critical Ecosystem Partnership Fund (CEPF)	International Livestock Research Institute (ILRI)
DISCONTOOLS	Office International des Epizooties (OIE) Performance of Veterinary Services
International Association for Ecology and Health	One Health Commission
EcoHealth Alliance	One Health Initiative
EcoHealth Network	One World, One Health Initiative
EcoHealth International Association for Ecology and Health	Swiss Tropical and Public Health Institute (Swiss TPH)
EcoHealth–One Health Resource Centers, Chiang Mai University and Gadja Mada University	Towards a Safer World
	Veterinarians Without Borders/Vétérinaires sans Frontières-Canada
Emerging Pandemic Threats (EPT) Program, USAID	World Bank Trust Fund for Avian Influenza
Epizone European Research Group	World Health Organization (WHO) Department of Food Safety and Zoonoses
Food and Agriculture Organization of the United Nations (FAO)	World Organization for Animal Health (OIE)
Global Early Warning System for Animal Diseases, Including Zoonoses (GLEWS)	World Small Animal Veterinary Association (WSAVA)
Global Framework for the Progressive Control of Transboundary Animal Diseases (GF-TADs)	

the next 20 years" declaring the "...need for sustained, well-coordinated, multi-sector, multi-disciplinary, community-based actions to address high impact disease threats that arise at the animal-human-environment interface."(UNSIC 2010).

2 Relevant Global and International One Health Endeavors in the Western Pacific Region

Numerous overlapping global and international initiatives from various consortia, donors, research institutes, and United Nations (UN) agencies are being implemented in the Western Pacific Region. While some initiatives have committed to improve coordination through systemic measures such as the *One World, One Health initiative* (FAO et al. 2008) and the *FAO-OIE-WHO collaboration concept note* on health risks at the human–animal interface (WHO et al. 2008), there is no overarching coordination of the multitude of activities being conducted in the region under a broad interpretation of One Health—this was emphasized at the recent Davos One Health summit with a major conclusion being the need to "intensify the collaboration and coordination between the leading and relevant...institutions in the broader One Health area" (Ammann 2012). In general, there are also no direct links with other global endeavors such as the Millennium Development Goals (UN Web Services Section 2010) and the Millennium Ecosystem Assessment (Millennium Ecosystem Assessment 2005).

Some of the global/international initiatives and organizations implementing these activities in the Western Pacific Region are listed below (Table 2). Only a selected number will be discussed in this chapter. This is an incomplete listing, but is illustrative of both the variety of work being addressed by various actors and institutions and the numerous (separate) networks operating in an environment of a broader One Health movement.

3 Selected Regional and Multicountry Steps Towards Operationalizing One Health

Western Pacific ministries of Health and Agriculture have had experience of being abruptly forced to work together in new ways to address new diseases. Not all interactions have been successful, and efforts to date have not yet fully embraced a One Health approach, as most stakeholders currently understand it; rather, most initiatives are continuing efforts to combat key EIDs. There are, however, a number of endeavors that illustrate the movement towards One Health. Most of these initiatives are at the regional rather than the national or community levels.

3.1 Regional Strategies

3.1.1 Asia Pacific Strategy for Emerging Diseases

The cornerstone of regional plans to confront EIDs is the *Asia Pacific Strategy for Emerging Diseases* (APSED) (World Health Organization 2010; World Health Organization, Western Pacific Region 2010). This is essentially a "health security" construct aiming to strengthen national systems to comply with the legal requirements of the International Health Regulations (2005) (World Health Organization 2005) and to improve national capacity to combat EIDs. The latest iteration of the strategy (2010) drew heavily on the lessons learned from the 2009 pandemic of influenza A/H1N1 and allows countries flexibility to decide how they can best achieve the vision of the eight areas of focus: (1) surveillance, risk assessment, and response; (2) laboratories; (3) zoonoses; (4) infection prevention and control; (5) risk communications; (6) public health emergency preparedness; (7) regional preparedness, alert, and response; and (8) monitoring and evaluation. The *Emerging Disease Surveillance and Response* unit of the WHO is responsible for assisting countries to implement APSED. APSED is not a One Health vision, however, lacking the synergies between all sectors whose activities impact on health.

3.1.2 Asia Pacific Economic Cooperation

Asia Pacific Economic Cooperation (APEC) considers emerging diseases to be of high importance because of their preventability and the substantial direct (e.g., treatment and hospitalizations) and indirect (e.g., lost time to work, trade sanctions) costs that such diseases have caused to their 21 members states in recent years. Since 1996, APEC has supported the APEC Emerging Infections Network (APEC EINet 2012), a network that seeks to gather and disseminate notifications of EIDs affecting APEC member states, foster collaborations among academic institutes, government, and business where they relate to EIDs, and enhance regional biopreparedness. This mechanism is useful for dialog between sectors beyond just the animal and human health sectors, although the degree of communication and idea sharing does not approach the *transdisciplinarity* advocated by most One Health proponents.

Nonetheless, APEC did fund the *Technology Foresight Project* (2006–2007) (The APEC Center for Technology Foresight National Science and Development Technology Agency 2008; Damrongchai et al. 2010), a succinct effort in transdisciplinarity that brought together a range of experts from policy makers and technology developers to virologists and economists to map the convergence of new technologies and the opportunities for their accelerated development in order to limit the human and financial impact of novel diseases. While narrowly focused on the technological aspects of disease prevention and control, and a project rather than an ongoing, inbuilt process, this work encompassed the development of new vaccines, treatments, diagnostics, models and simulations, and tracking strategies for people and animals.

APEC have since drafted a *One Health Action Plan* (Asia–Pacific Economic Cooperation 2011) setting out a common "vision" for member states to operationalize One Health approaches according to their capacities and level of engagement with the concept. The plan aims to strengthen cross-sectoral efforts at the political and leadership level, in teaching and training, in (government) functions to prevent, investigate, respond and control diseases, and across borders. The community is identified as a critical partner in disease prevention and control, and action to ensure the sustainability of cross-sectoral approaches is called for.

3.1.3 ASEAN

The Association of Southeast Asian Nations (ASEAN) has defined a *roadmap* to prevent, control, and eradicate highly pathogenic avian influenza (HPAI) and other highly pathogenic emerging diseases among member states by 2020 using a risk-based approach to address the major transmission pathways in each country (ASEAN Secretariat 2010a). The roadmap describes itself as a "translation" of the One Health approach to systematically eradicate HPAI, while simultaneously addressing other transboundary and zoonotic diseases. While the focus is on animal health and production, the advantages of engaging with multiple disciplines, multiple sectors, and multiple agencies are noted.

This is an encouraging output from ASEAN, but is one of the few documented instances of ASEAN activities related to One Health, either in progress or completed. Furthermore, the emphasis on HPAI rather than a broader One Health approach potentially misses an opportunity to embrace a wider notion of health including the role of wildlife, the integration of resources from various health and nonhealth authorities, as well as concrete plans for regular communication across health and related disciplines. ASEAN is in a unique position to be the premier institution in Asia coordinating, influencing, and even governing to some degree an integrated One Health approach for part of the Western Pacific Region. The HPAI roadmap is a step in the right direction but much remains to be done if ASEAN is to be a One Health leader. ASEAN's biggest challenge may be the reluctance of member nations to advise on what others should be doing. This is, however, a requirement for an integrated One Health network to be effective among the member states.

The *ASEAN Plus Three* EIDs *Programme* has improved joint country investigations of disease outbreaks and developed a regional risk communication strategy (ASEAN Secretariat 2010b). A new program funded by the Japanese Government is directed at improving laboratory capacity and networking (ASEAN Secretariat 2009), continuing a long and successful history of Japanese funding to develop diagnostic and research laboratory capabilities in the region.

3.1.4 The Food and Agriculture Organization of the United Nations

The *Food and Agriculture Organization (FAO) Regional Strategy for Highly Pathogenic Avian Influenza and other* EIDs *of Animals in Asia and the Pacific, 2010–2015* (Emergency Centre For Transboundary Animal Diseases 2010) outlines a common approach for dealing with endemic HPAI and for addressing emerging and re-emerging diseases. The strategy also aims to join up the fragmented support provided by various partners and donor agencies within the region. This is the latest in a series of initiatives led by FAO and its partners to combat HPAI since the first outbreaks in Southeast Asia, initiatives that were themselves preceded by other efforts founded in One Health concepts including the FAO Emergency Prevention System (EMPRES) and the Global Framework for the Progressive Control of Transboundary Animal Diseases (GF-TADs).

3.1.5 Donor Strategies

Most of the major Pacific Basin donors have made significant contributions to initiatives to address emerging diseases. The Public Health Agency of Canada leads the *Canada–Asia Regional Emerging Infectious Disease (CAREID) Project* aiming to strengthen the capacity of Cambodia, Laos PDR, the Philippines, and Vietnam to detect and respond to emerging diseases (Public Health Agency of Canada 2012). Similarly, the Australian Government's international development assistance agency has articulated a regional strategy for strengthening health systems to respond more generally to EIDs: the *Pandemics and* EIDs *Framework 2010–2015* (AusAID 2010). The *Asian Development Bank* has implemented a series of communicable diseases control projects along borders in the Greater Mekong region to improve community surveillance of endemic and epidemic diseases including EIDs (Asian Development Bank 2012). And the USAID *Emerging Pandemic Threats Program* (U.S. Agency for International Development 2010) operates globally with specific activities related to four project areas in Southeast Asia: wildlife pathogen detection, risk determination and reduction, outbreak response capacity, and institutionalization of a One Health approach. This last element is elaborated on in the next section (Academic Initiatives).

The European Union (EU) has also been active in supporting One Health initiatives through a range of endeavors. The flagship is the *EU Regional Highly Pathogenic Emerging Diseases (HPED) in Asia Programme* (2009–2013) (European Commission 2012) which spans two WHO regions, the Western Pacific and the South–East Asia Regional Offices. It aims to help ASEAN and the South Asian Association for Regional Cooperation (SAARC) to control, respond, and prepare for these diseases, and aligns with specific initiatives of OIE, FAO, and WHO via separate projects channeled through these three UN specialized agencies.

The European External Action Service recently published a comprehensive examination and summary of One Health case studies, many of which are active in the region, and a complementary database of One Health initiatives, studies, and actors (Hall and Coghlan 2011). This publication is well positioned to act as a guide in identifying individuals who can serve as One Health focal points in the region and to provide a starting point for operationalizing regional activities and networking in One Health.

It is important to note that the EU has commented in a number of fora that an approach to One Health needs to be positioned with consideration of societal needs. This "whole of society" approach to health hazards will require a wide-scale change in the attitudes and perspectives health professionals hold with regard to risk management.

3.2 Academic Initiatives

3.2.1 Southeast Asia One Health University Network

There are a number of existing networks of Southeast Asian Universities that encompass aspects of One Health such as the ASEAN University Network, the Asia Partnership on Emerging Infectious Disease Research, the Asia Pacific Academic Consortium for Public Health, the Asian Ecohealth Network, and the Southeast Asia Veterinary Schools Association. Through the RESPOND component of the *Emerging Pandemic Threats Program* (U.S. Agency for International Development 2010), USAID is supporting a new One Health academic collaboration, Southeast Asia One Health University Network (SEAOHUN) (Fenwick 2011), that brings together multiple faculties including schools of medicine, veterinary science, public health, and allied sciences from universities throughout the region. Cambodia, Laos, Indonesia, Malaysia, Thailand, and Vietnam have universities that belong to the network with China, Myanmar, and the Philippines to join in 2013. The network aims to develop transdisciplinary capacity to investigate and control outbreaks of emerging diseases and to build the evidence base for One Health approaches through research. This effort will define One Health competencies and develop a common regional approach to incorporating them into accredited education and professional in-service training.

3.3 Surveillance and Laboratory Initiatives

3.3.1 The Mekong Basin Disease Surveillance Initiative

The Mekong Basin Disease Surveillance Initiative (MBDS) (Mekong Basin Disease Surveillance 2007a) is a network established in 1999 to advance cooperative action among the six countries of the Mekong subregion to improve infectious disease surveillance and outbreak response. This aims to "reduce morbidity and mortality from communicable diseases, particularly amongst marginalized people living in the Mekong region" (Mekong Basin Disease Surveillance 2007b). From sharing of surveillance data from four border sites in 2003, the scope of the network has expanded through a second memorandum of understanding (MOU) in 2007 to the consideration of community-based surveillance, epidemiology capacity, information and communications technologies, risk communications, laboratory capacity, policy research, and extended cross-border cooperation. These seven new strategies will contribute to the development of national capacities identified in the International Health Regulations (2005) (World Health Organization 2005) to detect, investigate, report, and respond to public health threats. While not originally envisaged as a One Health activity and lacking some of the attributes of a One Health network, MBDS nevertheless provides a successful framework on which One Health approaches can be added or modeled.

3.4 Projects and Programs

3.4.1 Southeast Asia Foot and Mouth Disease Campaign

The Office International des Epizooties (OIE) coordinates the Southeast Asia Foot and Mouth Disease (SEAFMD) (OiE 2002) across eight ASEAN countries, a program recognized internationally as a model for regional coordination of animal disease control. Although foot-and-mouth disease (FMD) is not normally considered as a zoonotic disease (it rarely causes mild skin lesions in humans), the model stands as an example of an integrated effort among government agencies, international organizations, village communities, and donors all committed to controlling one disease. Individual national plans are harmonized with a regional strategy that has received high-level political commitment and that has adopted a progressive, long-term approach for the eradication of FMD. Close cooperation and the introduction of new techniques including zoning to roll back FMD in various parts of Southeast Asia including Malaysia and Thailand have contributed to the success of the program.

3.4.2 International Livestock Research Institute

Together with a large number of partners, International Livestock Research Institute (ILRI) is involved in a number of initiatives that could be deemed pertinent to One Health. *Ecohealth approaches to the better management of zoonotic emerging infectious diseases in the Southeast Asia Region* (*EcoZEID*) (Gilbert 2011) adopts a learning by doing approach in six countries aiming to demonstrate how capacity for research and disease control can be developed to address specific risks and impacts of EIDs. ILRI also manages the *Field Building Leadership Initiative* (*FBLI*)*: Advancing ecohealth in Southeast Asia* (China, Indonesia, Thailand, and Vietnam) (Tung DX 2011). This program combines research, capacity building through education and in-service training, and knowledge translation through connections to policy makers to design sustainable agricultural practices that result in improvements to human health, livelihoods, and environments.

3.4.3 Community-Based Avian Influenza Risk Reduction Program

CARE Australia implemented locally tailored community-level pilot projects to enhance disease surveillance and reduce risk behaviors related to avian influenza in Cambodia, Laos, Vietnam, and Cambodia during 2007–2009 (AusAID 2008). Although this program has concluded, the ensemble of projects elucidated some of the earliest lessons for operationalizing One Health in the Western Pacific Region: the importance of political, organizational, and community commitment to move lessons from pilots into systematic practice; sustained application of resources to stimulate lasting culture change; and the value of mixing multiple disciplines and agencies to overcome the Gordian knot of competing priorities in order to develop acceptable, effective solutions.

3.5 Research Initiatives

3.5.1 Asian Partnership on Emerging Infectious Disease Research and the Building Ecohealth Capacity in Asia Project

Supported by Canada's International Development Research Centre (IDRC), Asian Partnership on Emerging Infectious Disease Research (APEIR 2012) expanded an earlier network focused on research to fight avian influenza [The Asian Partnership for Avian Influenza Research (APAIR)]. The Chinese Academy of Sciences, the Thai Ministry of Public Health and the National Research Council, the Ministry of Science and Technology of Vietnam, and the Cambodian Ministry of Research and Technology and the Ministry of Health comprise the partnership to generate multidisciplinary research based on a broader ecohealth rather than One Health concept.

In the region, IDRC is also co-funding with the Australian Agency for International Development (AusAID) a related project on a smaller scale: the Building Ecohealth Capacity in Asia (BECA) project (Hall et al. 2012) which aims to increase involvement of researchers in Ecohealth and One Health initiatives. Although this is a relatively small project, it has been contributing to building a network of researchers working with several of the initiatives outlined in this chapter.

3.5.2 Ecohealth Emerging Infectious Diseases Research Initiative

Along similar lines, the Canadian and Australian Governments jointly fund the Ecohealth Emerging Infectious Diseases Research Initiative (Eco EID) (IDRC CRDI 2012), a multicountry project supporting research on how diseases emerge and spread in Southeast Asia and China, as well as developing research capacity and improving the translation of research into policy.

3.5.3 National Center of Competence in Research North–South

A rare nondisease focused approach branded with the One Health label, National Center of Competence in Research (NCCR) (National Center of Competence in Research North–South 2012) has mapped changing land use patterns and the transformation of agriculture in Lao PDR and Vietnam and are linking this with public health and economic impacts for small-scale farmers. NCCR is also documenting the health issues faced by internal migrants in Vietnam whose movements and changing employment have been triggered by rapid economic development. These activities demonstrate the potential breadth of the One Health approach in moving beyond traditional ideas of the boundaries of health.

3.5.4 Zoonotic Emergence Network, China and Malaysia

The majority of emerging disease "events" since 1940 have been zoonoses and the majority of these jumped from wildlife (Jones et al. 2008). Such viral spillovers have focused attention on interactions with wildlife and their environments; wild animals are also increasingly being farmed in Southeast Asia. A group of partners selected Malaysia because of Nipah virus and China because of SARS coronavirus to examine the risk of viral emergence among people regularly exposed to diverse animal species (hunters, indigenous people, and market workers) (EcoHealth Alliance 2012). Project partners include Ecohealth Alliance, the Global Viral Forecasting Initiative, the Malaysian Ministry of Health and Departments of Wildlife and National Parks and Veterinary Services, the Guangdong Entomological Institute and Centers for Disease Control, and the Chinese Institute of Zoology. This network brings together a range of animal and human health actors to specifically study aspects of the crucial intersection among animals, humans, and the environment that the One Health approach intends to address.

4 Selected National Level One Health Activities

Rather than overtly applying One Health approaches, national level planning has, with few exceptions, evolved along targeted planning for specific diseases coupled with some generic pandemic preparedness.

4.1 Cambodia

In Cambodia, the *National Committee for Disaster Management* (*NCDM*) has ultimate responsibility for dealing with emergencies of any nature and has played a key role in coordinating responses to HPAI. Specific plans for how ministries cooperate during emergencies have been outlined and align separate departmental plans (Sovann 2006). The bureaucratic arrangement of responses to zoonoses under an emergency structure is different from many countries in the region. In principle, however, a One Health approach includes the components of disaster risk reduction as expressed in the *Hyogo Framework for Action*, 2005–2015 (United Nations International Strategy for Disaster Reduction 2007).

4.2 China

The *Global Environmental Institute* is a Beijing-based nongovernmental organization that seeks to develop sustainable market-based models to untangle domestic environmental issues through engagement with local communities, government agencies, research groups, civil society, and the private sector (The Global Environmental Institute 2012). Unlike most initiatives in the Western Pacific Region, this organization is not being driven by concerns about specific infectious diseases and embraces a somewhat broader concept of One Health that intimately links with private enterprise. From a similar perspective, Kunming Medical University and the World Agroforestry Centre, Kunming has been developing projects and programs to address national ecohealth issues. Both institutions have played key roles in leading One Health and ecohealth research in China, particularly research in mountainous regions.

4.3 Lao PDR

The *National EID Coordination Office* of the Government of Laos has recently been established a Zoonosis Coordination Mechanism that enables collaborative action by the Ministry of Health, Agriculture, and Forestry to control zoonotic diseases (Lao Voices 2011).

4.4 Malaysia

Outbreaks of a new disease, Nipah virus, led to the formation of an *Interministerial Committee* for the control of zoonotic diseases directly linking human and veterinary health actors. Nipah virus provided a key case study of how disease incursions from wildlife can be amplified by human activities and rapidly spread in the absence of sensitive surveillance systems and rapid responses. Further research continues under the *Zoonotic Emergence Network (ZEN), China and Malaysia* (as discussed above). The Interministerial Committee has drafted an Infectious Disease Outbreak Rapid Response Manual (Ministry of Health Malaysia 2003).

4.5 Pacific Island Countries and Territories

The imperative to link-up animal and human health actors in Asia has been less pressing in the Pacific where the livestock sector is smaller and where HPAI has had only a limited impact. Low population density, the nature of market value chains in which livestock may be less likely to return to vendors, and fewer migratory bird flyways associated with HPAI may be other reasons for the slower development of One Health activities in the Pacific. Nevertheless, Pacific Island Countries and Territories have been a general source of concern for the region in that any weak link increases the regional vulnerability to emerging and re-emerging infectious diseases. Under the umbrella of international and regional programs such as GF-TADs (OIE Regional Representation for Asia and the Pacific 2012) on the animal health side and International Health Regulations (IHR) and APSED on the human health side, sectoral capacities have been gradually improving. Efforts to develop a regional One Health strategy, however, have not yet been realized. One Health approaches have obvious application in addressing endemic diseases of animals in some of the larger Pacific states, as well as efforts to conserve biodiversity.

4.6 Philippines

The Filipino Government has established an *Inter-Agency Committee on Zoonoses* composed of representatives from the Department of Health, Department of Agriculture, and Department of Environment and Natural Resources (Aquino III BS 2011).

4.7 Vietnam

Vietnam has been one of the countries worst affected by HPAI (A/H5N1) in terms of the impact on the formal and informal agricultural sectors and on human life. The government quickly developed a joint program run by the Ministry of Health and Ministry of Agriculture and Rural Development to address H5N1. This joint action culminated in a new strategy, *The Vietnam Integrated National Operational Program on Avian Influenza, Pandemic Preparedness And Emerging Infectious Diseases (AIPED), 2011–2015: Strengthening responses and improving prevention through a One Health approach* (Vietnam Ministry of Agriculture and Rural Development and Ministry of Health 2011). While still focusing on the elimination of H5N1, the strategy has adopted a risk-based approach to attending to the drivers of disease emergence to prevent a range of known and unknown communicable diseases. It involves government, nongovernment, community, and private actors. It remains to be seen how well this can be implemented. Nonetheless, this is one of the first incorporations of One Health principles in a national plan.

With the support of USAID, Vietnam has also been active in developing an academic network to support training and research in One Health. *The Vietnam One Health University Network (VOHUNET)* is part of the SEAOHUN.

5 Conclusion

The application of One Health in the Western Pacific Region is in an early phase with few concrete examples of successful operationalization; even from these few examples there appear to be many areas of duplication and lack of coordination. Nonetheless, serious attempts at articulating attributes of One Health considered important for the region have been made in strategies and documents at the regional level with serious commitment to implement One Health approaches. This illustrates the sharp shift in thinking about the components of disease control and preparedness that have come with the surfacing of new diseases; the need for broader input from numerous sectors and the involvement of communities are seen as essential to balance competing ideas and to generate creative, innovative responses. Relearning the age-old lesson that human action (and inaction) plays a fundamental role in disease emergence has renewed focus on the possibilities of prevention including prevention that reaches to tackle determinants far upstream. This, however, demands even greater levels of interaction and communication to manage complex human and natural ecosystems.

The Western Pacific Region is well positioned to exploit the opportunities that have emerged with recent, dramatic outbreaks of new diseases and to accelerate the development of better ways of managing both new and old infections for the local, regional, and global good. These opportunities are also opportunities for donors involved in health and agriculture and those committed more generally to socioeconomic development to join up siloed initiatives. And the vacuum of governance is yet another opportunity to establish a means of leadership, mentorship, and coordination in the region to reduce inefficiencies, link disconnected networks, improve understanding and knowledge transfer, and speed capacity development and preparedness planning.

References

Ammann WJ (2012) Chairman's summary remarks, GRF One Health summit 2012. In: Proceedings of global risk forum One Health summit Davos 2012 "One Health, One Planet, One Future", Davos, Switzerland

APEC EINet (2012) Public Health Informatics, University of Washington School of Public Health. https://www.cphi.washington.edu/projects/foundational-research/apec-einet. Accessed 13 June 2012

APEIR (2012) Asian partnership on emerging infectious disease research. http://www.apeiresearch.net/main.php. Accessed 12 April 2012

Aquino III BS (2011) Administrative order no. 10, April 11, 2011: creating the Philippine Inter-agency Committee on Zoonoses, defining its powers, functions, responsibilities, other related matters and providing funds thereof. Supreme Court E-Library. http://elibrary.judiciary.gov.ph/index10.php?doctype=Administrative%20Orders&docid=13267651001687696250. Accessed 12 April 2012

ASEAN Secretariat (2010a) Crossing borders and traversing bridges of partnerships—roadmap for an HPAI-Free ASEAN community by 2020. ASEAN Secretariat, Jakarta

ASEAN Secretariat (2010b) The official website of the Association of Southeast Asian Nations. Joint statement of the 4th ASEAN+3 Health Ministers Meeting Singapore, 23 July 2010. http://www.aseansec.org/24936.htm. Accessed 12 April 2012

Asian Development Bank (2012) Greater Mekong Subregion (GMS) regional communicable diseases control (CDC) project: Cambodia, Lao PDR, Viet Nam. http://www2.adb.org/Documents/IndigenousPeoples/VIE/38017-VIE-IPDP.asp. Accessed 12 April 2012

Asia–Pacific Economic Cooperation (2011) APEC One Health Action Plan: a framework to assist APEC economies strengthen cross-sectoral networks and functioning against the threat of emerging and zoonotic infectious diseases. http://www.zoonoses.csiro.au/Resource.aspx?id=24. Accessed 22 July 2012

AusAID (2008) Community-based avian influenza risk reduction program for the Mekong Region Phase 2: mid-term review report. www.ausaid.gov.au/Publications/Documents/cbairrp-mtr.pdf. Accessed 12 April 2012

AusAID (2010) Pandemics and emerging infectious diseases framework 2010–2015. Commonwealth of Australia, Canberra

Charron DF (2012) In: Charron DF (ed) Ecohealth research in practice. Innovative applications of an ecosystem approach to health. International Development Research Centre, Ottawa

Damrongchai N, Satangput P, Tegart G et al (2010) Future technology analysis for biosecurity and emerging infectious diseases in Asia–Pacific. Sci Public Policy 37(1):41–50

EcoHealth Alliance (2012) Zoonotic Emergence Network (ZEN), China and Malaysia. http://www.ecohealthalliance.org/health/22-zoonotic_emergence_network_zen_in_malaysia_china. Accessed 12 June 2012

Emergency Centre For Transboundary Animal Diseases, Regional office for Asia and the Pacific and the Food and Agriculture Organization of the United Nations (2010) FAO regional strategy for highly pathogenic avian influenza and emerging disease of animals in Asia and the Pacific 2010–2015. FAO, Rome. www.fao.org/docrep/012/i1474e/i1474e00.pdf. Accessed 12 April 2012

European Commission (2012) Development and cooperation — EuropeAID. Highly pathogenic emerging diseases (HPED) in Asia. http://ec.europa.eu/europeaid/where/asia/regional-cooperation/animal-human-health/hped_en.htm. Accessed 1 July 2012

FAO, OIE, WHO, UNSIC, UNICEF and World Bank (2008) Contributing to One World, One Health: a strategic framework for reducing risks of infectious diseases at the animal-human-ecosystems interface. FAO, OIE, WHO, UNSIC, UNICEF, World Bank. http://www.fao.org/docrep/011/aj137e/aj137e00.htm. Accessed 1 July 2012

Fenwick S (2011) Development of a One Health University Network in South East Asia. Presented to the progress meeting on ecosystem approaches to the better management of zoonotic emerging infectious diseases in the South East Asian region, Bangkok, 10–13 December 2011. Bangkok, DAI Thailand. http://www.slideshare.net/ILRI/development-of-aonehealthuniversitynetworkinseasia. Accessed 1 July, 2012

Gilbert J (2011) One Health approaches to prevent and control zoonoses. Presented to the Agriculture, Fisheries & Conservation Department (AFCD) workshop on "One Health: Past, Present, and Future". Hong Kong, 27 September 2011. http://www.slideshare.net/ILRI/one-health-approaches-to-prevent-and-control-zoonoses. Accessed 12 April 2012

Hall DC, Coghlan B (2011) Implementation of the One Health approach in Asia and Europe: how to set-up a common basis for action and exchange of experience. European Union: European External Action Service, Brussels. http://eeas.europa.eu/health/docs/index_en.htm. Accessed 1 July 2012

Hall DC, Nguyen-Viet H, Willyanto I et al (2012) Improving food safety in Asia through increased capacity in ecohealth. J Public Health Epidemiol (In Press)
IDRC CRDI (2012) Ecosystems and human health. Canada–Australia research partnership for the prevention of emerging diseases. http://www.idrc.ca/EN/Programs/Agriculture_and_the_Environment/Ecosystem_Approaches_to_Human_Health/Pages/Eco-EID.aspx. Accessed 12 June 2012
Jones KE, Patel NG, Levy MA et al (2008) Global trends in emerging infectious diseases. Nat 451:990–993
Lao Voices (2011) Laos discusses zoonotic disease co-ordination mechanism. http://laovoices.com/laos-discusses-zoonotic-disease-co-ordination-mechanism/. Accessed 12 June 2012
Mekong Basin Disease Surveillance (2007a) http://www.mbdsoffice.com/. Accessed 1 July 2012
Mekong Basin Disease Surveillance (2007b) http://www.mbdsoffice.com/aboutus.php. Accessed 1 July 2012
Millennium Ecosystem Assessment (2005) http://www.maweb.org/. Accessed 13 June 2012
Ministry of Health Malaysia, Disease Control Division (2003) Infectious diseases outbreak rapid response manual. http://jknns.moh.gov.my/doc/cdc/Infectious%20Diseases%20Outbreak-Rapid%20Response%20Manual.pdf. Accessed 12 April 2012
National Center of Competence in Research North–South (2012) Newsletters. No.8 February 2012. www.north-south.unibe.ch/content.php/page/id/251. Accessed 12 June 2012
OiE (2002) SEAFMD Campaign—Home Page. http://www.seafmd-rcu.oie.int/index.php. Accessed 12 April 2012
OiE Regional Representation for Asia and the Pacific (2012) Global framework for the progressive control of transboundary animal diseases (GF-TADs). http://www.rr-asia.oie.int/representation/programmes/programme_g/index.html. Accessed 12 April 2012
Population Reference Bureau (2011) World population data sheet. http://www.prb.org/DataFinder/Topic/Rankings.aspx?ind=30. Accessed 12 April 2012
Public Health Agency of Canada (2012) Canada–Asia regional emerging infectious disease (CAREID) project. http://www.phac-aspc.gc.ca/id-mi/careid-eng.php. Accessed 1 July 2012
Sovann R (2006) Mainstreaming disaster risk management into local and national development planning. Presented at the 6th meeting of the ADPC regional consultative committee on disaster management, 9–11 November, Kunming, PRChina. http://www.rccdm.net/index.php?option=com_docman&task=doc_view&gid=123&Itemid=215. Accessed 12 June 2012
The APEC Center for Technology Foresight National Science and Development Technology Agency (2008) Roadmapping converging technologies to combat emerging infectious diseases: a project of the Asia–Pacific Economic Cooperation (APEC) and the Industrial Science and Technology Working Group (ISTWG). APEC Secretariat, Singapore. http://publications.apec.org/publication-detail.php?pub_id=117. Accessed 13 June 2012
The Global Environmental Institute (2012) http://www.geichina.org/index.php?controller=Default&action=index. Accessed 12 June 2012
Tung DX (2011) Field building leadership initiative: advancing eco-health in Southeast Asia. Presented to the progress meeting on ecosystem approaches to the better management of zoonotic emerging infectious diseases in the South EastAsian region, Bangkok, 10–13 December 2011. http://www.slideshare.net/ILRI/field-building-leadership-initiative-advancing-ecohealth-in-southeast-asia. Accessed 12 April 2012
UN Web Services Section (2010) A gateway to the UN system's work on the MDGs. Department of Public Information, United Nations. http://www.un.org/millenniumgoals/. Accessed 13 June 2012
UNDP (2011) Human development reports. International human development indicators. http://hdr.undp.org/en/data/map/. Accessed 12 April 2012
United Nations International Strategy for Disaster Reduction (2007) Hyogo framework for action: building the resilience of nations and communities to disasters. 2005–2015, Geneva. http://www.unisdr.org/we/coordinate/hfa. Accessed 12 April 2012

UNSIC (2010) International ministerial conference on animal and pandemic influenza, Hanoi, 20–21 April 2010. Summary, achievements and next steps. Report prepared by UNSIC
U.S. Agency for International Development (2010) Emerging pandemic threats: program overview. http://pdf.usaid.gov/pdf_docs/PDACP822.pdf. Accessed 12 April 2012
Vietnam Ministry of Agriculture and Rural Development and Ministry of Health (2011) The Vietnam integrated national operational program on avian influenza, pandemic preparedness and emerging infectious diseases (AIPED), 2011–2015: strengthening responses and improving prevention through a One Health approach. Vietnam Ministry of Agriculture and Rural Development and Ministry of Health, Hanoi. http://www.avianinfluenza.org.vn/index.php?option=com_remository&Itemid=228&func=fileinfo&id=168. Accessed 1 July 2012
World Bank (2009) People, pathogens and our planet Vol 1: towards a One Health approach for controlling zoonotic diseases. World Bank, Washington
World Health Organization (2005) International health regulations (2005). http://www.who.int/ihr/en/. Accessed 12 April 2012
World Health Organization (2010) Securing our region's health: Asia Pacific strategy for emerging diseases. World Health Organization, Geneva. http://www.wpro.who.int/emerging_diseases/documents/SecuringsRegHealth15/en/index.html. Accessed 12 April 2012
World Health Organization, the Food and Agriculture Organization of the United Nations and the World Organisation for Animal Health (2008) Zoonotic diseases: a guide to establishing collaboration between animal and human health sectors at the country level. World Health Organization, Geneva
World Health Organization, Western Pacific Region (2010) Asia Pacific strategy for emerging diseases (APSED, 2010). http://www.wpro.who.int/emerging_diseases/APSED2010/en/index.html. Accessed 12 April 2012
World Health Organization, Western Pacific Region (2012). http://www.wpro.who.int/countries/en/index.html. Accessed 12 April 2012

One Health Approach in the South East Asia Region: Opportunities and Challenges

Gyanendra Gongal

Abstract The outbreaks of SARS, avian influenza, and Nipah virus in Asian countries clearly demonstrated that new highly infectious agents periodically emerge at the human–animal interface. The experiences of regional countries with prevention and control of avian influenza, SARS have reinforced the need for sustained, well-coordinated, multi-sector, multi-disciplinary, community-based actions to address emerging disease threats. 'One Health' is a cost-effective, sustainable, and practical approach to find solutions for problems which need holistic, multidisciplinary approaches, particularly in resource-constrained countries. While there is a growing recognition of One Health, it has to be translated from concept into actions through country level activities that are relevant for specific situations.

Contents

G. Gongal (✉)
Disease Surveillance and Epidemiology, WHO Regional Office for South East Asia, New Delhi 110 002, India
e-mail: Gongalg@SEARO.WHO.INT

Current Topics in Microbiology and Immunology (2013) 366: 113–122
DOI: 10.1007/82_2012_242

Published Online: 21 July 2012

1 Introduction

One Health is an international movement to promote a holistic multidisciplinary approach at the animal–human–ecosystem interfaces. One health is not a new idea or principle but it is simply recognition of the need to promote a culture of working together in a sustainable way, particularly in resource-constrained countries, to address health risks at the human–animal interface.

The outbreaks of SARS, avian influenza, and Nipah virus in Asian countries clearly demonstrated that new highly infectious and/or highly pathogenic agents periodically emerge at the human–animal interface, and will continue to emerge in the future. The unprecedented outbreaks of these diseases had serious impacts on travel, trade, and tourism, and it was realized that prevention and control of emerging infectious and high impact diseases required a holistic, multidisciplinary approach. The experiences of regional countries with avian influenza and pandemic (H1N1) 2009 have reinforced the need for sustained, well-coordinated, multi-sector, multi-disciplinary, community-based actions to address emerging disease threats that arise at the human–animal interface.

2 Background

The International Ministerial Conference on Avian and Pandemic Influenza (IMCAPI) held in New Delhi in December 2007 recognized that highly pathogenic avian influenza (HPAI) was deeply entrenched in several countries and the current state of veterinary services and preparedness levels in most countries of Asia and Africa posed a serious threat to the prevention and control of HPAI and other infectious diseases. Thus, the New Delhi conference appealed to the global community to begin to address broader issues around the emergence and spread of HPAI and other emerging infectious diseases through international partnerships (Press Information Bureau of the Government of India 2007). Each Government should encourage functional links between human and animal health systems, while investing in sustainable capacity for preventing and controlling high impact diseases in animals.

The New Delhi Road Map offered a valuable benchmark for the preparedness efforts of national authorities and proposed a convergence between animal and human health systems. The conference also requested international partners to develop a strategic framework for implementation of One Health and present it at the sixth IMCAPI held in Sharm Al Sheikh in October 2008. The Strategic Framework was the joint product of six major international organizations: the Food and Agriculture Organization (FAO), the World Health Organization (WHO), the World Organisation for Animal Health (OIE), the United Nations Children's Fund (UNICEF), the World Bank, and the United Nations System Influenza Coordination (UNSIC). The strategic framework for reducing risks of infectious diseases at the animal–human–ecosystem interfaces was presented at the ministerial conference. The document sets out six priority objectives for countries to consider, such as developing capacity in surveillance, promoting interagency and cross-sectoral partnerships, and ensuring functioning national emergency response capacity (Contributing to One World 2008). The application of a One Health approach aims not only to minimize the local and global impact of epidemics and pandemics caused by emerging infectious diseases, but also to look at more holistic approaches for solving health-related problems in line with International Health Regulations (2005).

The seventh International Ministerial Conference on Animal and Pandemic Influenza held in Hanoi in April 2010 reiterated the need to move forward the one health approach at country level (http://www.imcapi-hanoi-2010.org/home/en/). The scope, priority, and approaches may be different in public health and animal health institutions and therefore it will be necessary to promote integrated and/or coordinated approaches for the implementation of 'One Health'. It was recognized that animal health sector is weak in terms of surveillance and response for emerging and high impact diseases and therefore more investment will be required for strengthening animal health services in developing countries. Indeed, the veterinary public health service is rudimentary in most developing countries and it is estimated that USD 1.3 billion will be required annually to implement a One Health approach until 2020 (http://siteresources.worldbank.org/EXTAVIANFLU/Resources/3124440-1172616490974/Fifth_Global_Progress_Report_July_2010.pdf). The technical discussions held for operationalizing One Health, from ideas to action in Winnipeg (2009) and Stone Mountain (2010), have encouraged academics, donors, and partners to synthesize concrete action points for pushing 'One Health' movement to a higher level. A high level technical meeting, jointly organized by FAO, OIE, and WHO, was held in Cancun in November 2011 to address health risks at human–animal–ecosystem interfaces, and identified key elements, high priority technical actions, and related practical next steps for moving forward on intersectoral collaboration, coordination, and communication (World Bank 2010).

3 Ground Reality

One Health activity is a spontaneous movement—there is no consensus on definition of One Health, and even at the One Health Conference held at Melbourne in 2011, no clear definition was apparent. While the general concepts are now well accepted, how to implement the One Health concept is still not clearly understood (http://www.who.int/influenza/human_animal_interface/HLTM_human_animal_ecosystems_nov_ 2011).

Each stakeholder understands it in a different way, and there is a lack of coordination and collaboration among them. A number of international donors and partners are involved in promoting the One Health concept through pilot projects in Asian countries and it has definitely contributed to advocating a One Health approach among intellectuals and professional groups but it has not yet generated a political commitment or the involvement of the government sector.

Currently, the One Health movement has, to some extent, been driven by an attractive hypothesis of 'microbe hunting' in wildlife. It does not necessarily mean that it is a priority of the host country to implement the project of this nature, or that it will have a significant role in changing public health or animal health policy. However, discovering novel agents in wildlife does not imply that they will be associated with human or livestock diseases; only when sudden disease outbreaks or epidemics arise and they are shown to be the cause do we acknowledge their role, such as in filoviruses, SARS, Nipah, and pandemic influenza virus. Thus, it is often difficult to prove the significance of newly discovered and previously unrecognized pathogens.

The European Union, World Bank, USAID, Rockefeller Foundation, CDC Atlanta, Public Health Agency of Canada, and other partners are actively engaged in creating regional forums but there is a real need for coordination between all international partners. The major criticism is that they are all working in isolation without any coordination and interaction. Often they are competing with each other. As a result, many workshops, seminars, and meetings are organized in the name of One Health, but they are largely limited to talk fests. A few One Health networks that are operating in Asia are as follows:

- A One Health University Network in Southeast Asia supported by Epidemic Pandemic Threat Programme under USAID.
- *One Health Alliance of South Asia* (*OHASA*) supported by the EcoHealth Alliance.
- A network of One Health Hubs in the South Asian Region. The World Bank supported Massey University project to fight zoonotic diseases through the development of joint disease investigations and "One Health Hubs" to link with other specialists across South Asia. So far, 67 health professionals from India, Pakistan, Sri Lanka, Bangladesh, Afghanistan, and Nepal have been trained in epidemiology concepts as part of the university's Master degrees.

Interestingly, these networks have a common agenda but there has been no coordination or collaboration between them.

FAO, OIE, and WHO indicated their intention to work more closely together and with their respective sectors to address health risks at the human–animal–ecosystem interfaces through Tripartite Concept Note released in April 2010 (http://www.oie.int/fileadmin/Home/eng/Current_Scientific_Issues/docs/pdf/FINAL_CONCEPT_NOTE_Hanoi.pdf). These organizations have established a tripartite coordination mechanism at the regional level in Asia in 2011 and they are working together to develop a functional coordination mechanism between human and animal health sectors at country level through various joint activities.

4 Issues to Be Considered

4.1 Priority

Each sector has its own mandate, responsibility, priority, and constraints. The animal health sector has a prime responsibility to control economically important transboundary animal diseases which affect food animals and livestock production. Avian influenza, brucellosis, anthrax, salmonellosis may be areas of interest for animal health since they have a major impact on quality (Food safety) and quantity (Food security) of livestock products and public health. Similarly, rabies, plague, and leptospirosis are major zoonotic diseases of public health concern which are transmitted by dogs and rodents but they have little impact on livestock production or animal health. It is therefore important to define the priority diseases which generate common interests for collaborative work. Zoonoses, food safety, and antimicrobial resistance are priority areas for mutual cooperation between two sectors depending on technical capacity, level of economic development, and export potential of livestock products.

4.2 Institutional Capacity

There are gaps in both animal health and public health systems in most Asian countries which require assistance to bridge through international partnerships and concrete action plans. The comparative advantages of each sector should be taken into consideration such as good laboratory capacity in the veterinary sector and good epidemiological capacity in the public health sector, depending on country, and each sector can complement the other. All countries are trying to establish coordination mechanisms between the human and animal health sectors, and there are success stories within the region which should be highlighted. It may be politic to demand equitable access to funds for both human health and animal health sectors, since human health will get top priority for funding in any country irrespective of economic and development status, whereas the most practical and feasible idea would be to create a pooled fund to support the prevention and control of zoonoses by allocating the necessary resources for institutional development and technical capacity building.

4.3 Ownership

It is easy to say that we have to work together, but it is difficult to work together if there is no common understanding or mutual interest. Often it is apparent that One health is 'owned' by a particular professional group, and other professional groups may feel uncomfortable working with them. Therefore, there must be advocacy for political commitment at the highest level to support ownership across the professions. The One health initiative must come from community and must take into account local needs and the prevailing situation.

4.4 *Sustainable Development*

Although a One Health approach is focused on prevention and control of highly pathogenic, emerging, re-emerging, and high impact diseases of humans and animals, this approach may have a far-reaching vision for sustainable and ecologically friendly development activities. One classical example may be ensuring food security and food safety through the development of sustainable agriculture. Antimicrobial and agro-chemical substances are indiscriminately used to boost agricultural and livestock production to feed an ever-increasing human population and fast growing livestock and poultry farming, but they are responsible not only for depleting natural resources but also for microbial and chemical hazards to human, animal, plant life, and to the environment. Global warming and environmental degradation have been created by expansion of agricultural land, intensification of agricultural production system, deforestation, and industrialization. It has been realized that a holistic multidisciplinary approach is needed to mitigate the negative impacts of man-made disasters. Community involvement is a prerequisite for sustainability of the 'One Health' movement in resource poor countries.

The One health movement is gathering momentum in some countries which may serve as a good practice and modality for others. Some examples from countries have been presented as follows.

4.5 *Bangladesh*

There is a coordination mechanism for zoonoses control between human health and animal health sectors which was historically established for avian influenza prevention and control. Recent outbreaks of anthrax and Nipah virus have demanded better intersectoral collaboration and WHO and FAO have been supporting pilot projects and workshops to share information and identify collaborative activities. Both human health and animal health sectors including academic institutions are working together to promote 'One Health' approach for zoonoses prevention and control.

The three Ministries namely Ministry of Health and Family Welfare, Ministry of Fisheries and Livestock, and Ministry of Environment and Forest, with the support of UNICEF, FAO, and WHO have developed a Strategic Framework for operationalization of One Health approaches for prevention and control of emerging, re-emerging, and high impact diseases in Bangladesh. It is important that donors and partners respect the aspirations of local champions of One Health, and support various activities specified in the Strategic Framework. The successful implementation of the Bangladeshi model will inspire other countries to operationalize similar modality suitable to local needs.

4.6 Bhutan

There is a well-established coordination mechanism for prevention and control of zoonoses at national level and joint activities have been launched for avian influenza and rabies. The human health and animal health sectors have developed a project proposal for a One Health approach through joint activities, networking, and multidisciplinary research. There are some dedicated local champions for the One Health cause in a small country like Bhutan, which is encouraging. International partnerships will help to develop a unique indigenous model for operationalizing a One Health approach at the human–animal–ecosystem interfaces in Bhutan.

4.7 India

Presently, there is a Joint Monitoring Group at national level to coordinate avian influenza prevention and control activities. The Ministry of Health and Family Welfare is taking an initiative to establish a coordination mechanism for zoonoses prevention and control at state and district levels and for promoting collaboration among human health, animal health, and municipal bodies through joint training programmes. It has been agreed to expand the coordination mechanism for zoonoses prevention and control at state and district levels in the 12th Five Year Plan (2012–2016) using FAO/OIE/WHO Guidelines for establishing coordination between human and animal health sectors.

Five priority zoonotic diseases have been identified for collaboration between human health and animal health sectors, i.e., anthrax, brucellosis, leptospirosis, plague, and rabies. An inventory of laboratories capable for diagnosis of zoonotic diseases in India has been developed so that a particular medical or veterinary laboratory may serve as a center of excellence for a specific zoonotic disease. The National Center for Disease Control and WHO have developed a curriculum for joint training of medical and veterinary professionals on zoonoses prevention and control including intersectoral collaboration. The Indian Council of Medical Research and Indian Council of Agricultural Research have been working together to promote multidisciplinary research for zoonoses and food safety through the funding of joint research activities. They have decided to designate a nodal institution in human health and animal health sectors to promote joint research activities and to create a network of institutions. The Public Health Foundation in India has been involved in promoting the One Health concept in operational research and training.

4.8 Indonesia

There was an inter-ministerial coordination committee for avian influenza (KOMNAS) in Indonesia which is being reorganized as the National Zoonoses Commission to consider the growing threats from emerging infectious diseases and zoonoses. More ground work needs to be done to make the commission functional. There is a mechanism for information sharing during avian influenza surveillance and outbreak investigation at the district level. Intersectoral meetings are organized at the provincial level to discuss avian influenza, rabies, and other diseases of common interest. EU, USAID, and AUSAID are supporting avian influenza control in live bird markets, and rabies control activities, through FAO and WHO who are the catalysts for strengthening intersectoral coordination and collaboration.

WHO sponsored a regional rabies coordination meeting in Maumere in 2011 to discuss and finalize the roadmap for "East Nusa Tenggara Province Free from Rabies 2017". The meeting was attended by representatives from public health, animal health, security, local government as well as representatives of FAO, WHO, and UN. A multisectoral workplan was agreed at the end of meeting and an agreement was signed for its implementation.

4.9 Myanmar

A national coordination mechanism between the human health and animal health sectors was established for avian influenza and pandemic influenza preparedness and it was functional during avian influenza outbreaks in the past. WHO and FAO have supported Ministry of Health, Livestock Breeding, and Veterinary Department, respectively, for joint activities such as outbreak investigation, field epidemiology training, and information sharing. Since 2008, field epidemiology training has been jointly organized by Ministry of Health and Livestock Breeding and Veterinary Department for public health and veterinary professionals working at State, Division, and Township levels. Joint animal–human health sector technical meetings were organized to strengthen intersectoral collaboration. A National Zoonoses Workshop was held in March 2011 and identified five priority diseases, i.e.,avian influenza (H5N1), anthrax, rabies, leptospirosis, and plague. The workshop recommended that a technical working group be formed to move the One Health agenda forward, and a stakeholder meeting should be organized to develop a roadmap for operationalizing a 'One Health' approach, considering the country needs.

4.10 Sri Lanka

As in other Asian countries, the threat from highly pathogenic avian influenza was major reason behind collaboration between the human health and animal health sectors for avian and pandemic influenza preparedness in Sri Lanka in 2006, although Sri Lanka has been able to maintain freedom from avian influenza. The government is planning to use intersectoral coordination mechanisms for elimination of rabies and the control of leptospirosis in line with One Health approaches. The World Bank is providing grant assistance to Government of Sri Lanka to support One Health activity for next 5 years.

4.11 Thailand

It is fortunate that there are a number of international agencies and partners stationed in Bangkok who are supporting One Health initiatives at national, regional, and international levels. There are local champions of One Health in academic institutions, and in both the government and non-government sectors. One Health Training-of-Trainers workshops have been organized by the Ministry of Public Health in collaboration with national and international partners to strengthen One Health Epidemiological Teams at the provincial and district levels. There are several initiatives at different level to run multidisciplinary training programme including One Health Master's programme at university level. There is a high level of One Health awareness at policy and professional levels, and Thailand is hosting the second One Health conference in 2013.

5 Conclusion

Prevention and control of emerging infectious diseases is an international public good. 'One Health' is a cost-effective, sustainable, and practical approach to find solutions for problems which need holistic, multidisciplinary approaches, particularly in resource-constrained countries. We have to understand that everyone can contribute to promoting One Health by understanding the interaction and interconnectivity of the human–animal interface. While there is a growing recognition of One Health, it has to be translated from concept into actions through country level activities that are relevant for specific situations. Country level activities should be focused toward strengthening the infrastructure, good preparedness, and pre-emptive measures for responding to emerging diseases and other acute public health problems.

Some people have started to believe that the One Health concept is an illusion being limited to talk shows. Universities and certain sectors such as public health, natural resources, wildlife, agriculture, etc., at country level are enthusiastic about participating in the One Health mission, as witnessed by monthly (or more often) programs of workshops and seminars in collaboration with several international organizations. However, they are often designed for 'grant or budget hunting' rather than aiming to raise public awareness and participation. Most medical and veterinarian students are unaware of these developments, and public education is needed for added community awareness.

Political commitment by national governments is fundamental in promoting a One Health approach for responding to and managing zoonotic diseases, and should be supported through policy decisions. There is a need for institutional development to operationalize and sustain practical applications of One Health at ground level with the support by local champions who may be working with government, non-government organizations, and academic institutions.

He who does not understand the whole, is condemned to be reborn—*The Upanishad*

Acknowledgments I am grateful to Prof. Thiravat Hemachudha and other colleagues for critical thoughts on One Health movement which will definitely contribute in operationalization of One Health in right path in the best interest of participating countries.

References

Press Information Bureau of the Government of India (2007) "Vision and road map" released at the International Ministerial Conference on Avian and Pandemic Influenza, New Delhi; December 2007

Contributing to One World (2008) One health. http://un-influenza.org/files/OWOH_14Oct08.pdf Accessed 25 Jan 2012

http://www.imcapi-hanoi-2010.org/home/en/Accessed 25 Jan 2012

http://siteresources.worldbank.org/EXTAVIANFLU/Resources/3124440-1172616490974/Fifth_Global_Progress_Report_July_2010.pdf Accessed 20 Mar 2012

World Bank (2010) People, pathogens and our planet. Volume 1: towards a one health approach for controlling zoonotic diseases. Accessed 25 Jan 2012

http://www.who.int/influenza/human_animal_interface/HLTM_human_animal_ecosystems_nov_2011.pdf Accessed 25 Jan 2012

http://www.oie.int/fileadmin/Home/eng/Current_Scientific_Issues/docs/pdf/FINAL_CONCEPT_NOTE_Hanoi.pdf Accessed 25 Jan 2012

One Health in Mongolia

Zayat Batsukh, B. Tsolmon, Dashdavaa Otgonbaatar, Baatar Undraa, Adyadorj Dolgorkhand and Ochirpurev Ariuntuya

Abstract The Asia Pacific Strategy for Emerging Diseases (APSED) requires collaboration, consensus, and partnership across all the different actors and sectors involved in different aspects of emerging disease. Guided by APSED, Mongolia has established a functional coordination mechanism between the animal and human health sectors. Surveillance, information exchange and risk assessment, risk reduction, and coordinated response capacity and collaborative research have been identified as the four pillars of the zoonoses framework. Intersectoral collaboration has been clearly shown to be a crucial tool in the prevention and control of emerging zoonotic diseases. A "One Health" strategy has been implemented under the concept of 'Healthy animal-Healthy food-Healthy people'. An intersectoral coordination

Z. Batsukh (✉) · B. Tsolmon
Government Implementation Agency for Veterinary and Animal Breeding, MoFA and Light Industry, Peace avenue-16a, SPB-IX, Ulaanbaatar, Mongolia
e-mail: zbatsukh@mail.mn

D. Otgonbaatar
National Centre for Zoonotic Diseases, Ministry of Health, 20 Khoroo, Songinokhairkhan district, 18131 Ulaanbaatar-211137, Mongolia
e-mail: da.otgon@yahoo.com

B. Undraa
National Center for Infectious Diseases with Natural Foci, Central post office box 582 Ulaanbaatar, Mongolia
e-mail: undraa_56@yahoo.com

A. Dolgorkhand
National Center for Infectious Diseases with Natural Foci, 20 Khoroo, Songinokhairkhan district, 18131 Ulaanbaatar, Mongolia
e-mail: khandadya@yahoo.com

O. Ariuntuya
WHO, P.O. Box 46/78 Ulaanbaatar, Mongolia
e-mail: ochirpureva@wpro.who.int; atuya_o@yahoo.com

Current Topics in Microbiology and Immunology (2013) 366: 123–137 123
DOI: 10.1007/82_2012_253

Published Online: 14 October 2012

mechanism established between the veterinary and public health sectors has expanded its function to incorporate more work on food safety, emergency management, and effects of climate change on zoonotic diseases. Its membership includes the human health sector, the veterinary sector, the national emergency management agency, the environment sector, emergency management and inspection authorities, and the World Health Organization (WHO). The main outputs of the coordination mechanism have been strengthened surveillance and response activities and laboratory capacities. The coordination mechanism has also strengthened the surveillance and response capacity of neglected zoonotic diseases, such as brucellosis, anthrax, and tick-borne diseases. Through regular meetings and brainstorming sessions, both sectors have developed joint operational plans, a long-term risk reduction plan 2011–2015, initiated a prioritization exercise and risk assessment for 29 zoonotic diseases, and reviewed and revised standards, procedures, and communication strategies. In 2011, a list of experts on major zoonoses were identified from different sectors and formed into a taskforce to identify the focal points for rabies, brucellosis, and vector-borne diseases. As a result, disease control strategies are now linked to scientific research and epidemiological expertise.

Contents

1 Background

1.1 Country Profile

Mongolia is a landlocked country in East and Central Asia, situated between and bordering China and Russia, and with a population of 2.7 million as of 2011. The country has the lowest population density in the world, one person per 1.57 km^2. Mining and agriculture, and are the two main sectors of the Mongolian economy. For centuries, the Mongolians have been engaged in animal husbandry, raising horses, sheep, goats, cattle, and camels. Agriculture, primarily herding, is the traditional basis of the Mongolian economy, contributing about 20 % of GDP and providing 40 % of national employment. Livestock husbandry is the main economic pillar, vital for public good, and the significant source of export income.

Due to increasing urbanization and socioeconomic development of country in recent years, migration from rural to urban and suburban areas has been increasing. In 2010, only 36.7 % of the population resided in rural areas. Approximately, 30 % of the population is nomadic or seminomadic. Administratively, Mongolia is divided into 21 provinces, and the capital city, Ulaanbaatar.

1.2 Climate

Mongolia has an extreme continental climate with long, cold winters and short summers, during which most precipitation falls. The temperature is as low as −45 to −50° C in the winter and can reach 25° to 30 °C in the summer. Global climate change is believed to have had an influence on the climate; the annual average climate temperature has risen by 1.94 °C over the last 65 years, and in the last 30 years, the temperature has risen faster and the rainfall has decreased in Mongolian forest-steppe regions. Due to environmental and human impacts in the last few years many rivers, streams, and lakes have dried, pasture growth has decreased by 20–30 %, pasture plant species numbers have reduced and it has resulted in an increase in land degradation and desertification. Natural disasters such as drought, heavy snowfall, flood, snowstorms, windstorms, extreme cold and hot temperatures, and earthquakes recurrently occur throughout the year. Mongolia is very dependent on nature and climate due to its traditional nomadic lifestyle throughout four seasons of the year.

The large herder population has a greater chance of zoonotic infections. As the Mongolian economy is heavily reliant on herding and agriculture, the harsh winters and periodic droughts have adverse effects on livestock and agriculture, and also on the health status of the population.

1.3 Situation of Zoonotic Diseases

The livestock population was 36.3 millions as of 2011, down from 44.0 million at the end of 2009. Pig and poultry population are not prominent. Endemic zoonotic diseases such as brucellosis, anthrax, rabies, plague, and tick-borne diseases create important public health problems.

In recent years, endemic zoonoses have expanded and outbreaks of number of transboundary diseases have emerged in both animals and humans. Climate change and extreme weather conditions have had an adverse effect on biodiversity, distribution of animals, and microflora, which can lead to the emergence of zoonotic agents and create favorable conditions for disease outbreaks. Over 20 bacterial and viral and 18 parasitic zoonotic diseases were reported in animals. Six out of 15 diseases listed as transmissible diseases notifiable to the OIE were reported in Mongolia, and four diseases have a potential risk for further spread.

The significance of zoonoses is increasing due to improved animal husbandry practices, climate change, desertification, and developments in the mining sector. In spite of the progress achieved, anthrax, brucellosis, tick-borne diseases, and rabies still constitute a threat to human health and welfare.

2 Coordinating Mechanisms Between Animal and Human Health Sectors

The Asia Pacific Strategy on Emerging Diseases (APSED) recognizes the importance of close multisectoral cooperation for the prevention and control of zoonoses. With the support of World Health Organization (WHO), the Intersectoral Coordination Committee on Zoonoses was officially established in Mongolia in February, 2010, although many collaborative activities had already been undertaken since 2006. The Committee is chaired by either the Vice-Minister of Health or the Vice-Minister of Food and Agriculture and Light Industry, alternating between the two positions annually, and the membership includes representatives from the Ministry of Health (MoH), Veterinary and Animal Breeding Agency of Ministry of Food and Agriculture and Light Industry (MoFALI), National Emergency Management Agency (NEMA), Ministry of Nature and Environment, General Agency for Specialized Inspection, and the WHO.

The overall vision of the Coordination Committee is to have “strong human and animal health sectors, together with emergency response and national inspection agencies working in partnership toward the attainment of a healthier community”. The Coordination Committee has responsibility for developing joint policy on the prevention and control of priority zoonotic diseases; for approving action plans produced by a technical working group; for making recommendations on risk assessment, early warning and response activities during outbreaks; for reviewing and revising zoonotic diseases standard operational procedures (SOPs) and guidelines to reflect intersectoral collaboration; for providing methodological assistance to improve the capacity of professional institutions at the national and subnational level; for coordinating cooperation among different sectors in carrying out early detection and response functions; and for monitoring and evaluating overall zoonotic disease prevention and control. The Director-General of the National Centre for Zoonotic Diseases in the MoH serves as secretariat, and is responsible for routine coordination and management.

Before the establishment of the Coordination Committee, MoH and MoFALI developed a written Memorandum of Understanding (MoU) to conduct joint surveys on zoonotic diseases in 2007–2009. Both sectors exchanged annual statistical reports and conducted joint serological surveys. The results of the survey helped define the distribution of major zoonoses which are important to both animal and human health. The surveys identified new diseases in Mongolia, such as tick-borne encephalitis, West Nile fever, Lyme disease, rickettsia, and Q fever. The joint survey promoted collaboration between two sectors. The new diseases have been added to the list of notifiable diseases to reflect current threats. However, most of the activities were aimed at gathering information about zoonotic pathogens only. Notable changes observed in the two sectors during the survey were transferred by the joint task to surveillance with ongoing and systematic collection of information in order to define the extent of disease problem, and to disseminate this information to improve public health awareness, early warning, diagnosis, prevention, and control.

The first meeting of the Intersectoral Coordination Committee took place in March 2010, and was attended by its members, the secretariat, the technical working group and evaluation team, as well as by representatives from WHO and FAO. The outcome of the meeting was discussion of the draft joint operational plan. The first activity was to map existing capacity and surveillance systems, and response and risk reduction measures in both the animal and human health sectors. Based on the results of this assessment, an operational plan of action was developed to address the gaps and to improve zoonose control strategies.

Quarterly meetings have been held and priorities set for actions and interventions. Regular meetings between veterinary and public health professionals proved to be an important activity to improve and stimulate intersectoral cooperation. During times of emergencies, both sides communicated frequently and joint technical working group meetings were conducted. A good example of this is the brainstorming joint response review meeting of veterinary and human health authorities in September 2010 following the outbreak of anthrax in animals and humans. All meetings are organized in cooperation with the WHO and other international organizations. The cost of organizing joint meetings and conferences was paid back by the harmonization of legislation, joint planning, and sharing of resources. This included sharing information and surveillance data and cooperation at the local level in outbreak response. This cooperation has been tested during real time outbreaks and the lessons learned from those exercises used to improve the rapid response measures.

The coordination committee organized the first national conference on zoonoses in June 2010. The participants were professionals from both the human and veterinary sectors at national and subnational levels. This was the first ever joint meeting between two sectors at a professional level. The meeting reviewed results of joint assessment on existing capacity and system for surveillance and response in the following areas:

- Human resources
- Response capacity
- Information and surveillance
- Laboratory
- Logistics and supplies.

After the National conference, the intersectoral coordination mechanism was formally set up at all levels in Mongolia. At the community level, social awareness, public education, and media play an important role. It has also enabled the use of better risk communication and health education strategies at the community level. Risk communication and promotion of programs directed primarily at occupational risk groups and school children were implemented with assistance from local government. At the national level, the coordination mechanism was aimed at improving information exchange, expertise sharing, mutual technical support, and harmonization of legislation. In 2011, a joint strategy for long-term risk reduction of priority zoonotic diseases for 2011–2015 was developed by the Ministries of Health and of Food and Agriculture.

3 Information Sharing, Surveillance, Risk Assessment, and Risk Reduction

3.1 Prioritization Exercise

The Intersectoral coordinating committee on zoonoses carried out a prioritization exercise and risk assessment of 29 zoonotic diseases in January 2011. These included endemic zoonoses reported in humans, zoonoses reported in animals, vector-borne diseases, and diseases at risk of being imported. A total of 16 zoonoses were identified that are important for both animal and human health sectors. The technical working group that consisted of veterinary, public health, laboratory, research institute, and academic personnel held a series of discussions and conducted detailed risk assessments. WHO's prioritization tool as well as other countries' methodologies and tools were adopted for this prioritization exercise. The priority diseases, namely, plague, avian influenza, anthrax, brucellosis, rabies, tickborne encephalitis, echinococcosis, and tularemia were defined as diseases that required a coordinated surveillance and response. Endemic diseases like brucellosis and anthrax, which have been listed by WHO as "neglected" were identified as priority diseases by MoH and MoFA. The exercise specially defined malaria, dengue fever, glanders, toxoplasmosis, West Nile fever, Japanese encephalitis, hemorrhagic fever with renal syndrome, and cryptosporidiosis as diseases that should be targeted for collaborative research.

3.2 *Sharing of Surveillance Data*

The coordination committee developed SOPs for information sharing, surveillance, and response for the priority diseases such as avian and pandemic influenza, anthrax, tick-borne diseases, rabies, brucellosis, plague, and some parasitic diseases. The veterinary and health sectors routinely cross-notify and exchange information, based on the SOPs. In addition to surveillance data, both sectors should exchange outbreak information within 24 h, and laboratory data and event information (immunization, cluster of cases, livestock abortion, sudden death of animals, survey results, food-borne disease) on a monthly basis. Weekly disease information has been shared with MoH, MoFA, WHO, FAO, and other partners through an electronic newsletter since March 2010.

3.3 *Brucellosis Control in Mongolia*

Mongolia has one of the highest incidences of human brucellosis in the world. National brucellosis surveillance was established in the 1950, and a test-and-slaughter strategy commenced in 1960. The Government implemented a vaccination strategy from 1973 to 1983. As a result, the prevalence of animal brucellosis has decreased from 10 to 0.5 %. However, in the 1990s human brucellosis re-emerged following transition to free market economy, collapse of systems that were responsible to public health issues and lack of resources to continue surveillance accordingly. In 2000, a new vaccination strategy was introduced with the aim of eradicating the disease by 2010, but attempts to control the disease have been unsuccessful because of inconsistent strategies with respect to vaccination of livestock and the detection and elimination of infected animals from the herd.

The seroprevalence of brucellosis in humans, livestock, and dogs was investigated as a pilot project in Sukhbaatar and Zavkhan province with support from Swiss Development Agency. The results of the study by veterinary and medical epidemiologists served as a baseline for assessing and monitoring the effectiveness of a conjunctival vaccination campaign in 2010. In addition, the conjunctival vaccine campaign has assisted the development of new strategy for national brucellosis control and for livestock export.

Despite the increase in the number of registered animal brucellosis cases, the MoH did not report an increase in the number of human brucellosis. In Mongolia, the disease incidence is largely unknown because many cases are missed due to a lack of diagnostic facilities at the subnational level. Only 2–3 % of cases of acute human brucellosis are reported, and it is estimated that less than one in 40 cases are reported indicating a significant under-reporting. Animal sector surveillance data helped the human health sector to review surveillance and laboratory practice to improve reporting. Brucellosis is identified as one of priority zoonoses for both animal and human health sector. In 2011, animal and human sector have started baseline prevalence survey. Over 200,000 serum samples from five major species of animals and 2,333 serum samples from human were collected and laboratory investigation were carried out, following OIE recommendations. A mass vaccination campaign has been implemented with the aim of controlling and eradicating animal Brucellosis by 2020. The country was divided into three sectors and 14.7 million animals were vaccinated in 2011 in 1st sector, with a future plan to vaccinate animals in remaining two sectors, and then to provide annual vaccination of newborns.

3.4 Joint Risk Assessment and Risk Reduction

In response to growing burden of anthrax in the Mongolia, a technical working group has developed a strategy for the prevention and control of human and animal anthrax. This is the first risk reduction disease strategy that has been prepared with involvement of human, animal, emergence management, inspection agency, food safety and intelligence authorities, and with international partners. The strategy has been based on global best practice and experience gained over the past 30 years of responding to outbreaks as well as sporadic cases of anthrax. A GIS-based risk map has been developed for anthrax to provide a common platform. In addition, a joint technical working group has been established with professionals from the Institute of Veterinary Medicine, the National Centre for Zoonotic Diseases, the Central Veterinary Diagnostic Laboratory to act as a professional advisory, and technical implementation body to develop methodological recommendations and policy documents for approval by relevant authorities.

In response to increasing numbers of rabies cases in wildlife, the veterinary and public health sectors have combined with local government over the past 2 years to conduct community education and awareness activities in schools, workplaces, and among the general population. On World Rabies Day 2011, the MoH organized a rabies awareness and prevention campaign and conducted training for healthcare workers, veterinarians, school doctors. The MoH also distributed brochures and posters for children, parents, and dog owners on rabies prevention, and video spots and cartoons were produced and broadcasted by media. The veterinary sector also initiated dog vaccination, and stray street dogs were destroyed in four districts.

An avian influenza surveillance program has been established in wild birds in order to provide an early warning system and to improve the existing surveillance network. The surveillance team consisted of representatives from the veterinary, health, environment, inspection, and other related institutions, and was a good example of multisectoral cooperation.

The two human and animal sectors have developed an epidemiological atlas of zoonotic diseases in Mongolia, 2011. The atlas contains approximately 50 maps that illustrate the distribution of major or rare and neglected zoonotic diseases. Every map contains key information about the infectious agent including: ICD-10 code, epidemiology, epizootiology, climate data, vegetation, transmission, incubation period, clinical findings, therapeutic options, and key references. In addition, the atlas includes population density, livestock density, antibiotic use, immunization coverage, and other relevant factors and will be regularly updated. It will be made available online by 2012. The use of GIS tools and geo-referenced, subnational level epidemiological data allowed the production of maps that improve spatial quality of previous maps. It was shown that diseases such as brucellosis, glanders, and bovine leucosis in animals have been introduced into previously unaffected areas by cattle movement. The atlas will lay the basis for novel, evidence-based methodologies to estimate the population at risk and burden of disease, ultimately leading to more targeted interventions. The atlas has also helped to streamline field data collection.

4 Coordinated Response to Emerging Zoonoses

Joint risk assessment and investigations have been conducted after cross-notification of outbreaks of foot-and-mouth disease, Newcastle disease, human and animal anthrax, rabies, and avian influenza in wild birds.

During outbreaks of anthrax, a rapid response team consisting of veterinarians, medical epidemiologist, inspectors and emergency officers, implemented quarantine and movement restrictions, and developed risk maps using GIS. Animal vaccinations, enhanced surveillance in the food market, and health education and communication activities has led to effective outbreak response. The subclinical, gastrointestinal form of anthrax was identified for the first time by the rapid response team.

Existing rapid response infrastructure has been improved into multisectoral joint rapid response teams that operate at the district and provincial levels; rapid response teams have been trained and established in 21 provinces.

Working together has made it possible to prevent zoonotic diseases, not merely to react to them once they have occurred. Laboratory integration, surveillance activities, and recognition of the importance of risk assessment have also increased.

5 Laboratory Cooperation

Under the APSED framework, communication and cooperation of veterinary and human health laboratories have increased significantly in the last 3 years. Laboratories share information, experience, diagnostic kits, laboratory specimens and lab equipment for surveillance, response, and research activities. Health laboratories have benefited from more advanced laboratory resources of veterinary laboratories, including personnel. During an unusual outbreak of human anthrax in 2011, the veterinary laboratory assisted in validating results and undertook confirmation tests. Subnational veterinary laboratories in all 21 provinces have been equipped with PCR equipment and reagents.

The veterinary laboratory also supported laboratory diagnosis of a rabies outbreak in Uvurkhangai province and in an unusual anthrax outbreak in Khovd province. Following annual serological surveys, the analysis of the laboratory findings was carried out jointly by laboratory staff from the veterinary and health laboratories, and the methodologies used in both sectors were reviewed and experiences shared.

As a result of human and animal sector collaboration, the diagnostic capacity of human health laboratories has been improved significantly. New advanced methods and techniques for isolation, identification, and confirmation of zoonotic viral and parasitic pathogens have been introduced at the national level. A number of commercially available diagnostic kits have been introduced for diagnosis at the NRCIDNF and the number of diseases diagnosed by molecular assays has increased to 17. Serological and molecular diagnostic tools have become available for the diagnosis of tick-borne encephalitis, Lyme disease, and Rickettsia which had previously been diagnosed only by clinical presentation. However, Hantavirus, West Nile virus, Japanese encephalitis virus, Crimean Congo hemorrhagic fever virus, dengue virus, and many others cannot be diagnosed due to technical limitations, and thus the true burden and epidemiology of these diseases in Mongolia is still unknown.

In addition to the collaboration with veterinary laboratories, training in advanced countries is seen as important for increase capacity at the laboratory diagnostic level. Since 2010, over 30 professionals have been trained in laboratory biosafety in Russian Federation, Kazakhstan, People's Republic of China, Germany, and Japan. Approximately 23 % of the trained lab professionals were from provincial veterinary and medical diagnostic laboratories.

As a result of collaborative molecular biology research with foreign colleagues from various countries including Russia, China, the USA, Germany, and Japan, various techniques such as Clustered Regularly Interspaced Short Palindromic Repeats (CRISPR), duplex polymerase chain reaction (PCR), variable number tandem repeats (VNTR), multiple loci VNTR analysis (MLVA), have been introduced to research and diagnostic laboratories for animal and human diseases, and have determined unique and specific genes of *Y.pestis, B.anthracis,* rabies virus, tick-borne encephalitis virus, and some species of Rickettsia. In addition,

Hantavirus, West Nile virus, Anaplasmosis, Erlichiosis, and Toxoplasmosis were newly identified using these techniques.

Several complications still exist that constrain sharing of resources between human and animal diagnostic laboratories and the biggest challenge for the Intersectoral Coordination Committee on Zoonoses will be to change the legal and ethical environment.

Mongolia is planning to establish a laboratory network between public health, clinical, veterinary, and food laboratories in 2012–2013.

6 Risk Communication

Lessons learned from managing previous outbreaks highlighted the importance of advocacy and public education. A communication and behavior change strategy was reviewed by the Coordinating Committee meeting in 2010. It emphasizes the need for advocacy and a public education campaign targeted at high-risk groups. A proactive approach in building effective communication with media was also stressed. Endemic zoonoses such as plague, anthrax, and vector-borne diseases occur regularly due to a lack of public awareness, and there is a high infectivity rate of brucellosis among herdsmen and veterinarians. Unsafe cultural traditions are widespread among the general population, such as consumption of raw milk, undercooked sheep liver, and sour cream made from raw milk. Public health education programs need to be aimed at specific community groups, school children, and occupational groups, taking into account culture, beliefs, traditions, educational level, social status, occupation, and age. An involvement of community and local government in health education through health education in schools and in the workplace has proved to be effective. Health messages on how to prevent infection with tick-borne diseases and the production of leaflets and posters were distributed before the tick season. In addition, a monthly press conference has been initiated by the MoH to ensure important public health messages are widely disseminated; the first press conference held on March 2011 advocated a One World, One Health approach to public health.

Regular awareness programs are conducted by State Veterinary and Animal Breeding Department, Institute of Veterinary Medicine, and the MoH through TV programs, brochures, video spots, cartoons for children, and press conferences.

Training materials and courses for risk reduction measures and interventions were developed for anthrax, plague, tick-borne diseases, brucellosis, and avian influenza collaboratively by animal and human health sectors. Joint staff training activities and short training courses on mosquito biology and surveillance, risk assessment of common zoonotic diseases, data management, database design, vector-borne diseases have been conducted for medical and zoonotic epidemiologists, biologists, laboratory staff, and meteorologists.

7 Collaborative Research

Tick-borne diseases such as tick-borne encephalitis, Lyme disease, and rickettsia are a growing concern in Mongolia, as their prevalence continues to increase with expansion into new areas. Pastoral animal husbandry, climate change, desertification, development of mining sector, new tick species, and vector distribution in Mongolia combine to create an important public health problem. To mitigate these risks, a Korean International Cooperation Agency (KOICA) funded project has supported vector surveillance, climatic monitoring and community education to high-risk population. This initiative is multisectoral, and is bringing together people with different backgrounds and sectors. At the regional level, emerging diseases surveillance and response (ESR) and malaria, vector-borne and parasitic disease (MVD) units are working together.

Climate change studies are complex and require multisectoral collaboration. Building on the achievements of the intersectoral coordination mechanism, a comprehensive surveillance system for vector-borne diseases has been established. Surveillance procedures have been developed for anaplasmosis, Q fever, tickborne encephalitis, tickborne boreliosis, rickettsia, and erlichiosis. Tick distribution and species are monitored in relation with microclimate and human infections. Erlichiosis and anaplasmosis, toxoplasmosis, and Crimean Congo hemorrhagic fever infections were identified for the first time in humans, and *Anaplasmosis platys* was identified for the first time in ticks. The veterinary laboratory is undertaking genetic studies on ticks.

Correlation of infected tick density with variations in human incidence and climate determinants has helped to identify factors associated with disease transmission. Risk maps on tick prevalence, density, biotype, climate data, and vegetation has provided useful public health information for early warning. Increased risk communication and staff training has resulted in improved protective behavior of the nomadic population.

8 International Partnership

The National Center for Zoonotic Diseases has established good collaboration and partnerships with many international organizations and institutions from various countries including China, Kazakhstan, Russia, Japan, Switzerland, the USA, and Germany. Epidemiologists interested in zoonoses have been cooperating with Chinese Academy of Inspection and Quarantine since 2007 on collaborative research directed at understanding the natural foci and the conditions affecting disease incidence each side of the border of both countries. This collaborative research has also enhanced laboratory capacity, including a substantial donation of virology laboratory equipment to the NCIDNF by the Chinese Academy of Inspection and Quarantine. The laboratory will be basis for conducting cross-border surveillance, on-the-job training of laboratory staff, and confirmation of events and diseases of public health importance.

The NCIDNF conducts collaborative research on plague and tick-borne diseases with the Bundeswehr Institute of Microbiology of Munich. Both institutions carry out annual joint field investigations and expeditions. The results from these studies have been published and presented at an international zoonoses conference held in Mongolia. Extensive research and cross border surveillance of bacterial, parasitic, and viral diseases have been conducted in collaboration with Gamalei Institute of Epidemiology and Microbiology, and natural foci of leptospirosis, cryptosporidiosis, and toxoplasmosis have been detected for the first time in Mongolia.

Studies on the molecular biology of plague, tick-borne diseases, and other emerging diseases have been undertaken by veterinary and public health specialists with colleagues from the University of Florida. An important part of the collaboration with the University of Florida is a 'One Health' training program which started in 2011, and which attracted a number of staff members from the Institute of Veterinary Medicine and the MoH. It is hoped that the course may attract the US and international students. The curriculum will include studies in environmental health, modern laboratory techniques, epidemiology, biostatistics, food safety, climate change, GIS, toxicology, and zoonotic infections research.

9 Challenges and Lessons Learned

APSED has facilitated an intersectoral coordination mechanism between human health and other sectors. However, although ongoing risk assessments are conducted during outbreaks, there has been no comprehensive cross-sectoral risk assessment for all priority zoonoses. Evidence-based decision making and response, and utilization of risk assessment findings, need to be further improved. It has also been realized that an enabling legal environment is critical for effective control of zoonoses. The annual intersectoral simulation exercise has been a useful way to review response capability, and to update and revise the coordinated response guidelines. Subnational level planning and information sharing between veterinary and health epidemiologists, however, is still weak. At the local level, the involvement of the veterinary health departments is crucial for effective monitoring of instances of zoonotic disease in wild and domestic animals. There is also need to improve both the health laboratory capacity and in epidemiological capacity in the animal sector. During the annual review meeting in 2011, the need for developing and implementing a common monitoring and evaluation framework was highlighted, and poor coordination and confusion over roles and responsibilities among veterinary, health and inspection agencies on food safety, and import and export control need to be addressed.

Financial contribution is crucial for the success of zoonoses control in the country, so that effort from MOFALI and MOH is requested to have more efficient way to raise the fund and harmonize international donor recourse, by drawing attention of potential donors for the activity in the zoonoses field.

We believe a good foundation has been established for a coordination mechanism between the veterinary and public health sectors, and the generic capacity for zoonoses control and prevention has improved considerably. In addition, the zoonoses coordination framework has attracted more resources from international partners and allowed pooling of resources. Thus while an important process has started, there is still much to do to reduce the risk of zoonotic diseases in Mongolia.

Editorial addition

Asia Pacific Strategy for Emerging Diseases (APSED) and its role in responding to zoonotic disease threats.

John S Mackenzie,

Curtin University, Perth, and the Burnet Institute, Melbourne

APSED was developed in 2005 as a joint initiative by the South-East Asian (SEARO) and Western Pacific Regional Offices (WPRO) of the WHO to meet the challenges of emerging diseases that pose serious threats to regional and global health security (WHO 2005). APSED provided a common strategic framework for countries and areas of the two regions to strengthen their capacity to manage and respond to emerging diseases including epidemic-prone diseases, and to develop the capability to comply with the core capacity requirements of the new International Health Regulations (2005). It had the support of all 48 regional countries (11 in SEARO and 37 in WPRO), and thus represented countries with a combined population of 3.4 billion people, more than half of the world's population.

The development of APSED was greatly influenced by several major emerging zoonotic disease events in the Asia Pacific region, and especially by the emergence of severe acute respiratory syndrome (SARS) and highly pathogenic avian influenza H5N1 (HPAI), as well as the initial outbreak and continued recurrences of Nipah virus. During the first 5 years of the Strategy, the two regions experienced a number of infectious disease threats including the establishment of HPAI as an endemic disease, the rapid global spread of pandemic influenza H1N1 2009, and a large number of other acute events with significant public health impact. Taken together, these provided important lessons in pandemic response and demonstrated the need to further strengthen public health emergency preparedness and improve monitoring and evaluation.

APSED (2005) recognized that many emerging diseases were zoonoses, and that an important component of the Strategy was the development of plans to detect, manage, and respond to infectious diseases at the human–animal interface. During the first 5 years of the Strategy, a guide was developed in collaboration with colleagues from the World Organization for Animal Health (OIE) and the Food and Agriculture Organization of the United Nations (FAO) entitled 'Zoonotic Diseases: A Guide to Establishing Collaboration between Animal and Human Health Sectors at the Country Level' to assist countries with their planning (WHO 2008).

Thus, considerable progress was made in the two regions toward strengthening core capacities needed to prevent, detect and respond to threats posed by emerging diseases, and has provided a good foundation for expanding the scope of APSED. This led to a Biregional Consultation to explore how to take the Strategy forward for the next 5 years, resulting in the development of APSED (2010) (WHO 2011). The new Strategy has expanded to eight focus areas, including zoonoses, with a strong statement recognizing the importance of zoonotic diseases and with an undertaking to continue working in collaboration with FAO and OIE and other partners … 'to contribute to the concept of "One Health"', and acknowledging that reducing the risk of transmission of zoonotic diseases requires close collaboration between and links with the food safety, environment, and wildlife sectors. It also states that the experience and lessons learned with HPAI (H5N1) provide a good foundation to consolidate and strengthen national and regional coordination mechanisms for surveillance information-sharing and coordinated responses by human and animal health sectors.

In response to the Strategy, a number of countries in the regions have developed plans to coordinate and collaborate between their human and animal sectors, and in some instances, also their environmental sectors, through a 'One Health' approach. Mongolia is one such example, and the description of their plans and activities clearly demonstrate how they are building a sustainable and collaborative approach toward managing zoonotic diseases, and developing the capacity to diagnose and respond to new emerging disease threats—a good example of operationalising 'One Health' at the national level. Other examples are given in the chapters by Dr G Gongal and Dr B Coughlan.

References

International Health Regulations (2005) World Health Organization, second edition, 2008, Switzerland

WHO (2005) Asia Pacific strategy for emerging diseases. World Health Organization Regional Office for South-East Asia, New Delhi, and the Regional Office for the Western Pacific, Manila

WHO (2008) Zoonotic diseases: a guide to establishing collaboration between animal and human health sectors at the country level. World Health Organization Regional Office for South-East Asia, New Delhi, and the Regional Office for the Western Pacific, Manila

WHO (2011). Asia Pacific strategy for emerging diseases, 2010. World Health Organization Regional Office for South East Asia, New Delhi, and the Regional Office for the Western Pacific, Manila

Part III
One Health New Technologies, New Approaches and How to Implement

Climate Change and Human Health: A One Health Approach

Jonathan A. Patz and Micah B. Hahn

Abstract Climate change adds complexity and uncertainty to human health issues such as emerging infectious diseases, food security, and national sustainability planning that intensify the importance of interdisciplinary and collaborative research. Collaboration between veterinary, medical, and public health professionals to understand the ecological interactions and reactions to flux in a system can facilitate clearer understanding of climate change impacts on environmental, animal, and human health. Here we present a brief introduction to climate science and projections for the next century and a review of current knowledge on the impacts of climate-driven environmental change on human health. We then turn to the links between ecological and evolutionary responses to climate change and health. The literature on climate impacts on biological systems is rich in both content and historical data, but the connections between these changes and human health is less understood. We discuss five mechanisms by which climate changes impacts on biological systems will be felt by the human population: Modifications in Vector, Reservoir, and Pathogen Lifecycles; Diseases of Domestic and Wild Animals and Plants; Disruption of Synchrony Between Interacting Species; Trophic Cascades; and Alteration or Destruction of Habitat. Each species responds to environmental changes differently, and in order to predict the movement of disease through ecosystems, we have to rely on expertise from the fields of veterinary, medical, and public health, and these health professionals must take into account the dynamic nature of ecosystems in a changing climate.

J. A. Patz (✉) · M. B. Hahn
Nelson Institute, Center for Sustainability and the Global Environment (SAGE), University of Wisconsin, 1710 University Avenue, Madison, WI 53726, USA
e-mail: Patz@wisc.edu

M. B. Hahn
e-mail: mbhahn@wisc.edu

Current Topics in Microbiology and Immunology (2013) 366: 141–171
DOI: 10.1007/82_2012_274

Published Online: 18 November 2012

Contents

Both benefits and challenges of a One Health approach when addressing emerging infectious diseases (Sect. 2), food security (Sect. 3), and national sustainability planning (Sect. 4) have been addressed elsewhere in this volume. Integration of the impact of climate change into these systems adds complexity and uncertainty that intensifies the importance of interdisciplinary and collaborative research. Awareness of links between ecological processes and diseases may be of value if anticipated changes in global climate—and subsequent changes in ecosystem structures and functioning—are considered simultaneously. As One Health practitioners, we emphasize the importance of understanding interactions among the health of humans, animals, and the environment. Each responds to climate change in a different way—across varying temporal and spatial scales, mechanisms, and levels of magnitude. Piecing together the web of interconnections within an ecosystem is difficult, and climate change introduces a dynamic element to the puzzle that creates a perpetually shifting target. Collaboration among veterinary, medical, and public health professionals to understand the ecological interactions and reactions to flux in a system can facilitate clearer understanding of climate change impacts on environmental, animal, and human health. We begin with a brief overview of the scientific background, and then examine several case studies of climate change impacts on ecosystem and animal health that have demonstrated links to human health as well.

1 What is Climate Change?

Climate change, whether due to natural variability or resulting from human activity, depends on the overall energy budget of the planet, the balance between incoming (solar) shortwave radiation and outgoing longwave radiation. This balance is affected by the Earth's atmosphere, in much the same way that a greenhouse (or a car's windshield on a hot day) allows sunlight to enter and traps heat (infrared) energy inside. An atmosphere that retains more heat, because it has higher levels of so-called greenhouse gasses, will lead to higher average surface temperatures.

A definitive source of information on climate change is the work of the United Nations Intergovernmental Panel on Climate Change (IPCC), which was established by the World Meteorological Organization (WMO) and the United Nations Environment Programme (UNEP) in 1988. Since 1990, the IPCC has conducted international assessments of the current scientific work on climate change, the potential impacts of this change, and various prevention options at approximately 5-year intervals. This international body includes many outstanding scientists who represent multiple sectors, and its reports are viewed as the most authoritative assessments on the subject. Much of the information on climate science in this chapter is drawn from IPCC reports (Solomon et al. 2007).

1.1 Greenhouse Gasses

The composition of the Earth's atmosphere has changed since preindustrial times. Beginning approximately in the mid-1800s, changes include increases in atmospheric levels of carbon dioxide (CO_2), methane (CH_4), and nitrous oxide (N_2O) that far exceed any changes occurring in the preceding 10,000 years. Historical levels of these greenhouse gasses are known from analyses of air trapped in bubbles in Antarctic ice cores (Etheridge et al. 1998; Gulluk et al. 1998). For example, the concentration of CO_2, the most significant greenhouse gas, has risen by approximately 35 %, from about 280 parts per million by volume (ppmv) in the late eighteenth century to about 380 ppmv at present. Higher greenhouse gas concentrations have contributed to warming the Earth in an effect called positive radiative forcing, by absorbing and re-emitting infrared radiation toward the lower atmosphere and the Earth's surface (Fig. 1 summarizes the principal components of radiative forcing). Figure 2 depicts the temperature changes across the globe since 1900 related to natural and anthropogenic forcings.

1.2 A Warming Earth: From Past to Future

Long-term climate change can be observed as a signal standing out against a background of natural climate variability (Fig. 2). To help in determining the meaning of this signal, we need historical climate data to measure natural variability. Because instrument records are available only for the recent past (not quite 150 years), previous climates must be deduced from paleoclimatic records such as tree rings, pollen series, faunal and floral abundances in deep-sea cores, isotope

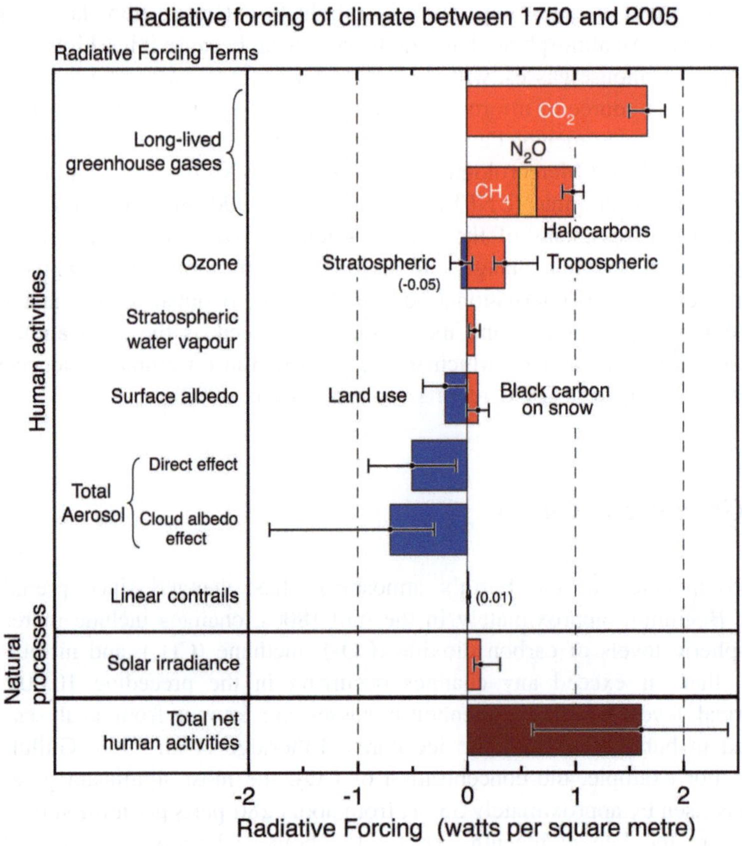

Fig. 1 Components of radiative forcing. Global mean radiative forcings (RF) and their 90 % confidence intervals in 2005 for various agents and mechanisms. Errors for CH_4, N_2O and halocarbons have been combined. The net anthropogenic radiative forcing and its range are also shown. Reproduced with permission from Solomon et al. (2007, FAQ 2.1, Figure 2)

analyses of coral and ice cores, and diaries and other documentary evidence. Results of these analyses show that average North American temperatures in the mid- to late twentieth century appear to have been warmer than during any similar period in the last five centuries and likely the highest in at least the past 1,300 years (Solomon et al. 2007). The increasing temperature trend is accelerating rapidly. From 1906 to 2005, global average temperature rose by 0.74 °C. According to the IPCC, by 2100 health-relevant weather extremes will be very likely (Table 1). The rate of change is faster now than in any period in the last 1,000 years.

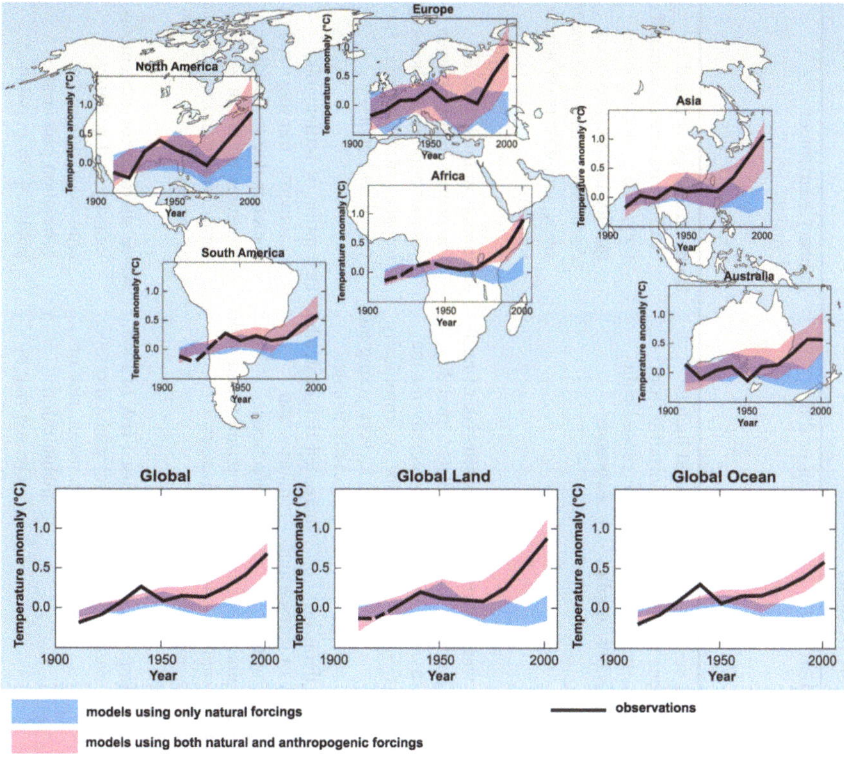

Fig. 2 Temperature changes due to natural and anthropogenic forcings. Comparison of observed continental- and global-scale changes in surface temperature with results simulated by climate models using either natural or both natural and anthropogenic forcings. Decadal averages of observations are shown for the period 1906–2005 (black line) plotted against the centre of the decade and relative to the corresponding average for the 1901–1950. Lines are dashed where spatial coverage is less than 50 %. Blue shaded bands show the 5–95 % range for 19 simulations from five climate models using only the natural forcings due to solar activity and volcanoes. Red shaded bands show the 5–95 % range for 58 simulations from 14 climate models using both natural and anthropogenic forcings. Reproduced with permission from Solomon et al. 2007, FAQ 9.2, Figure 1)

1.3 Earth System Changes

Although warming is the average effect across the Earth's surface, changing temperatures are only part of the story. Higher temperatures evaporate soil moisture more quickly (thus severe droughts), while warm air can hold more moisture and result in heavy rains; such "hydrologic extremes" (floods and droughts) are very much a part of climate change scenarios and therefore of substantial concern to public health professionals. In addition, Arctic and Antarctic ice are melting, thereby releasing vast amounts of water into the oceans, raising

Table 1 Projected Earth system changes

Phenomenon[a] and direction of trend	Likelihood of future trends based on projections for 21st century using SRES scenarios	Examples of major projected impacts by sector			
		Agriculture, forestry and ecosystems	Water resources	Human health	Industry, settlement and society
Over most land areas, warmer and fewer cold days and nights, warmer and more frequent hot days and nights	Virtually certain[b]	Increased yields in colder environments; decreased yields in warmer environments; increased insect outbreaks	Effects on water resources relying on snowmelt; effects on some water supplies	Reduced human mortality from decreased cold exposure	Reduced energy demand for heating; increased demand for cooling; declining air quality in cities; reduced disruption to transport due to snow, ice; effects on winter tourism
Warm spells/heat waves. Frequency increases over most land areas	Very likely	Reduced yields in warmer regions due to heat stress; increased danger of wildfire	Increased water demand; water quality problems, e.g. algal blooms	Increased risk of heat-related mortality, especially for the elderly, chronically sick, very young and socially isolated	Reduction in quality of life for people in warm areas without appropriate housing; impacts on the elderly, very young and poor
Heavy precipitation events. Frequency increases over most areas	Very likely	Damage to crops; soil erosion, inability to cultivate land due to waterlogging of soils	Adverse effects on quality of surface and groundwater; contamination of water supply; water scarcity may be relieved	Increased risk of deaths, injuries and infectious, respiratory and skin diseases	Disruption of settlements, commerce, transport and societies due to flooding: pressures on urban and rural infrastructures; loss of property
Area affected by drought increases	Likely	Land degradation; lower yields/crop damage and failure; increased livestock deaths; increased risk of wildfire	More widespread water stress	Increased risk of food and water shortage; increased risk of malnutrition; increased risk of water- and food-borne diseases	Water shortage for settlements, industry and societies; reduced hydropower generation potentials; potential for population migration

(continued)

Table 1 (continued)

Phenomenon[a] and direction of trend	Likelihood of future trends based on projections for 21st century using SRES scenarios	Examples of major projected impacts by sector			
		Agriculture, forestry and ecosystems	Water resources	Human health	Industry, settlement and society
Intense tropical cyclone activity increases	Likely	Damage to crops; windthrow (uprooting) of trees; damage to coral reefs	Power outages causing disruption of public water supply	Increased risk of deaths, injuries, water- and food- borne diseases; post-traumatic stress disorders	Disruption by flood and high winds; withdrawal of risk coverage in vulnerable areas by private insurers; potential for population migrations; loss of property
Increased incidence of extreme high sea level (excludes tsunamis)[c]	Likely[d]	Salinisation of irrigation water, estuaries and fresh-water systems	Decreased fresh- water availability due to saltwater intrusion	Increased risk of deaths and injuries by drowning in floods; migration-related health effects	Costs of coastal protection versus costs of land-use relocation; potential for movement of populations and infrastructure; also see tropical cyclones above

Examples of possible impacts of climate change due to changes in extreme weather and climate events, based on projections to the mid- to late 21st century. These do not take into account any changes or developments in adaptive capacity. The likelihood estimates in column two relate to the phenomena listed in column one.

[a] See Working Group I Table 3.7 for further details regarding definitions

[b] Warming of the most extreme days and nights each year

[c] Extreme high sea level depends on average sea level and on regional weather systems. It is defined as the highest 1% of hourly values of observed sea level at a stationfor a given reference period

[d] In all scenarios, the projected global average sea level at 2100 is higher than in the reference period. The effect of changes in regional weather systems on sea levelextremes has not been assessed

Reprinted with permission from Solomon et al. (2007), Table SPM.3, Geneva, Switzerland their levels (from melting land-based glaciers) and potentially altering the flow of their currents. The weather patterns that result from these and other changes vary greatly from place to place, and over short periods of time, emphasize the importance of climate variability. For these reasons, the term climate change is more accurate than global warming and is the accepted term for these phenomena.

Accordingly, the accelerating temperature changes noted above have been correlated with the Earth system changes. Since 1961, sea level has risen on average by approximately 2 mm per year (Solomon et al. 2007), and snow cover and glaciers have diminished in both hemispheres. Most striking is the extent to which the Arctic ice cap has melted in the past 30 years. These trends are forecast to continue. According to the IPCC, in 90 years sea level will rise between 18 and 59 cm. Extremes of the hydrologic cycle (e.g., floods and droughts) are also expected to accompany global warming trends.

2 Impacts of Climate-Driven Environmental Change on Human Health

The relationships between human health and environmental changes due to climate change have been reviewed extensively elsewhere (McMichael et al. 2006; Patz et al. 2000, 2005) and are summarized in Fig. 3. Direct effects from a warming climate have been shown in several studies on the correlation between heat waves and excess mortality (Curriero et al. 2002). Natural disasters such as floods, droughts, and intense storms have claimed millions of lives during the past two decades, and affected many more physically, mentally, or through the loss of property or livelihoods (International Federation of Red Cross 1998). Further, the IPCC's midrange sea level rise projections (a 40-cm rise by the 2080s) will put 200 million people at risk for a range of health problems including displacement, salt water intrusion into fresh water aquifers, or disruption of stormwater drainage and sewage disposal (IPCC 2007). Warmer temperatures are likely to affect air quality through changes in ozone concentrations, a known pulmonary irritant associated with pneumonia, chronic obstructive pulmonary disease, asthma, and premature mortality (Ebi and McGregor 2008). In addition, extensive research suggests that climate change will influence aeroallergens and related human allergic disorders through changes in pollen season (Beggs 2004; Ziska et al. 2011). Nutrition and food security will also be affected through changes in crop yields, unreliability of supplies, and impacts on prices (Battisti and Naylor 2009; Schmidhuber and Tubiello 2007).

Food- and waterborne diseases are likely to become a greater problem as climate changes. For example, flooding can contaminate drinking or recreational water with pollution from sewage lines or agricultural fields (Lipp et al. 2001; Thomas et al. 2006). Heavy rainfall can overwhelm sewage systems and treatment plants, which then discharge excess wastewater directly into surface water bodies (Patz et al. 2008).

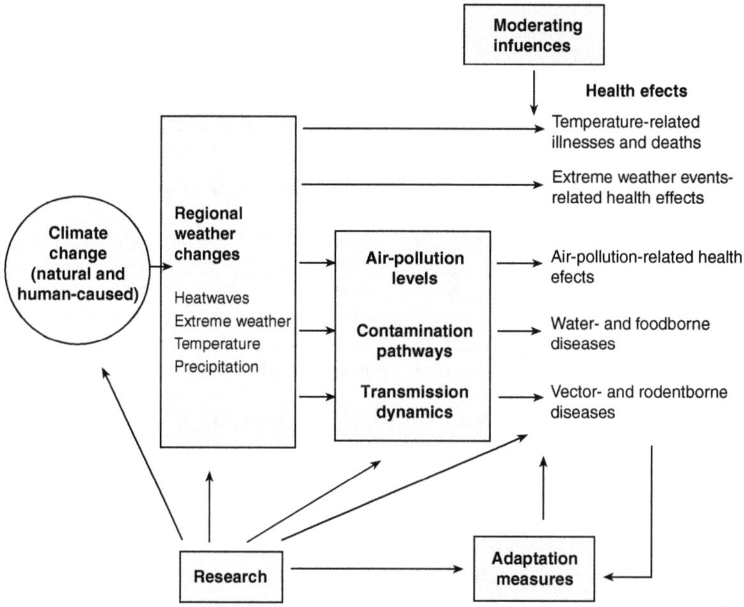

Fig. 3 Potential health impacts of climate variability and change. Moderating influences include non-climate factors that affect climate-related health outcomes, such as population growth and demographic change, standards of living, access to health care, improvements in health care, and public health infrastructure. Adaptation measures include actions to reduce risks of adverse health outcomes, such as vaccination programs, disease surveillance, monitoring, use of protective technologies (e.g., air conditioning, pesticides, water filtration/treatment), use of climate forecasts and development of weather warning systems, emergency management and disaster preparedness programs and public education. Reproduced with permission from Patz et al. (2000)

Disease outbreaks from most waterborne pathogens are distinctly seasonal and cluster in key watersheds (Curriero et al. 2001). There is strong evidence that links incidence of waterborne outbreaks from pathogens such as *Cryptosporidium* (MacKenzie et al. 1994), *Escherichia coli* 0157:H7 (Hrudey et al. 2003), and *Campylobacter jejuni* (Hrudey et al. 2003) following heavy rains. Storm events of >3 inches of rainfall within 24 h can overwhelm combined sewer systems and lead to an overflow that contaminates recreational and drinking water sources (Patz et al. 2008). For example, levels of *E. coli* in channels leading from Milwaukee to Lake Michigan can be up to ten times higher in areas where there are no sewage overflows (Fig. 4). Climate change is anticipated to increase the frequency of these occurrences. Regional climate models, for example for the Great Lakes area of the United States, show a 50–120 % increase in sewage overflow events by the end of this century (Patz et al. 2008). This will pose increased hazards to drinking and recreational water quality. In Peru, childhood diarrheal rates increased 200 %

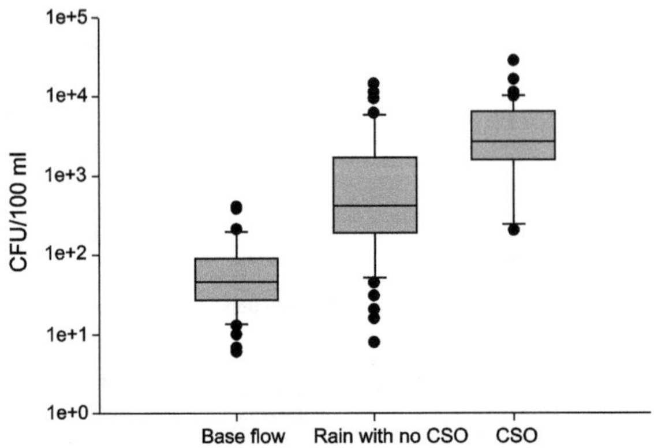

Fig. 4 Levels of *E. coli* in the Milwaukee Estuary and rain events in channels leading to Lake Michigan with and without combined sewer overflow (CSO) systems. Levels of E. coli in the Milwaukee estuary, which discharges to Lake Michigan, 2001–2007, during base flow (n = 46); following rain events with no CSO (n = 70); and following CSO events (n = 54). Boxes indicate 75 % of values, with median values drawn in each. Whiskers are 95 % of values and outliers are shown as closed circles. There were significant differences in E. coli levels following rainfall and CSOs compared to base flow ($p \leqq 0.05$). Reproduced with permission from Patz et al. (2008)

during the 1997–1998 El Nino episode, likely due to higher survival time of diarrhea-causing pathogens in combination with an increase in warm-weather behavior such as higher demand for water and less conscientious hygiene practices (Checkley et al. 2000) (Fig. 5). The worldwide average for diarrheal diseases in the future is projected to rise 20 % for the period 2040–2069 and 29 % for 2070–2099 (Kolstad and Johansson 2011).

The relationship between foodborne outbreaks and temperature has been shown for several pathogens and in a variety of geographic settings (Bentham and Langford 2001; Lake et al. 2009; Zhang et al. 2007). For example, (D'Souza et al. 2004) showed that increases in salmonellosis notifications in five Australian cities were related to a rise in the mean temperature in the previous month. Heat contributed to an estimated 30 % of cases of salmonellosis in much of continental Europe, especially when temperatures exceeded a threshold of 6 °C above average (Kovats et al. 2004). A recent re-evaluation of foodborne illness over time in England and Wales confirmed the correlation with temperature in the current and previous week. The study discusses the importance of lowering pathogen loads in livestock through vaccinations of chicken flocks, limits on antibiotic use in cattle to retard development of resistant strains, and improvement of hygiene practices in abattoirs as methods to help curtail foodborne outbreaks (Lake et al. 2009).

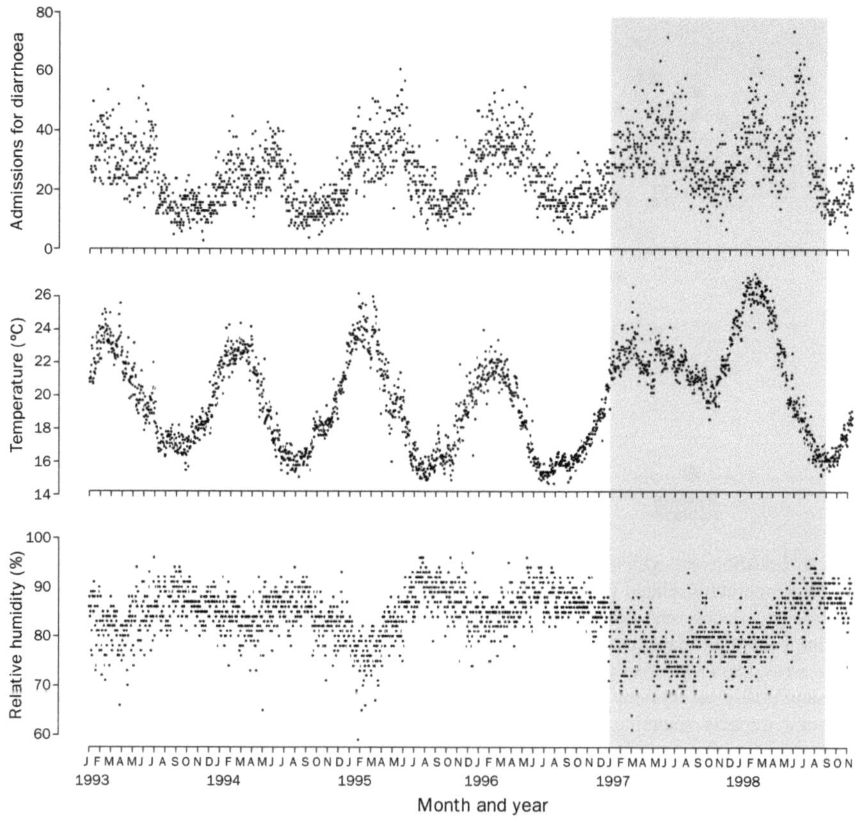

Fig. 5 Daily time series between Jan 1, 1993 and Nov 15, 1998, for admissions for diarrhoea, mean ambient temperature and relative humidity in Lima, Peru. Shaded area = 1997–98 El Niño event. Reproduced with permission from Checkley et al. (2000)

3 Drawing Connections: Discovering the Links Between Ecological and Evolutionary Responses to Climate Change and Human Health

Research on climate change impacts on biological systems is a rich field of study with origins in the late eighteenth and early nineteenth centuries when (Bumpus 1899) documented the effects of an extreme winter storm on the selection of body size in house sparrows *(Passer domesticus)* and (Grinnell 1917) observed the role of temperature in defining the geographic range of many species (Parmesan 2006). More recent studies on geographic range shifts have benefited from long-term observational records of many species taken by dedicated naturalists, for example, Aldo Leopold's observations on the timing of spring events on a Wisconsin farm (Bradley et al. 1999; Parmesan 2006). The IPCC has documented long-term

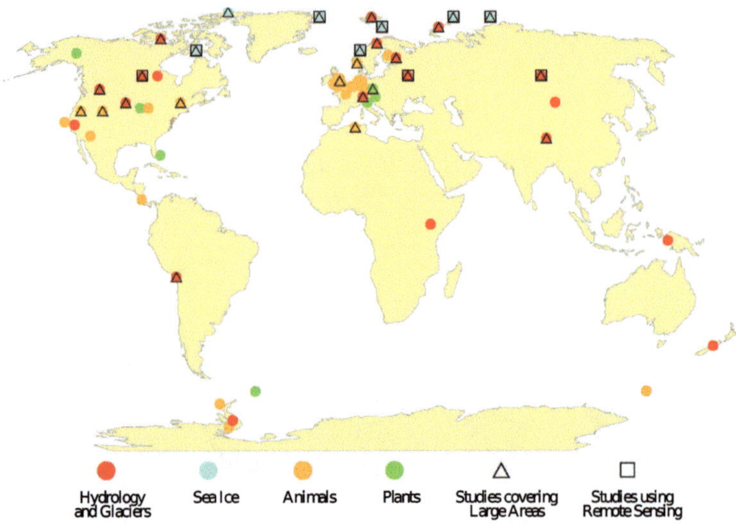

Fig. 6 Locations at which systematic long-term studies meet stringent criteria documenting recent temperature-related regional climate change impacts on physical and biological systems. Hydrology, glacial retreat and sea-ice data represent decadal to century trends. Terrestrial and marine ecosystem data represent trends of at least two decades. Remote-sensing studies cover large areas. Data are for single or multiple impacts that are consistent with known mechanisms of physical/biological system responses to observed regional temperature-related changes. For reported impacts spanning large areas, a representative location on the map was selected. Reproduced with permission from Kovats et al. (2001, Figure TS-11)

studies of biologic responses to climate change on every continent (Fig. 6). Less well studied are the cascading relationships between climatic shifts, adaptation, alteration, or extinction of biota, and their resultant impacts on human health. One reason for this lack of continuity may be the dearth of multisystem, multidisciplinary research studies on climate change impacts, and the propensity to simplify complex issues by carving out small, more manageable pieces of the puzzle for investigation. Although highly detailed, system-specific information on biological impacts of climate change are necessary, integration of these biologic studies into a larger ecological web of effects will make them more relevant and useful than treating them in isolation. It is here that we begin to draw connections between species responses to climate change and human health and demonstrate areas in which biology, veterinary medicine, and public health overlap.

3.1 Modifications in Vector, Reservoir, and Pathogen Lifecycles

Vector-borne diseases are one of the most studied health challenges of climate change (Table 2). These are diseases that are transferred to humans or among humans by an arthropod vector. The potential relationship between climate and

Table 2 Selected examples of climatic factors influencing the transmission and distribution of vector-borne diseases. Reproduced with permission from Gage et al. (2008)

Disease (Causative Agent)	Vector(s)	Relevant climatic factors	Effects of climatic variability or climate change
Parasitic Vector-Borne Diseases			
Malaria (*Plasmodium vivax*, *P. falciparum*)	Mosquitoes	Temperature, rainfall, humidity, El Nino-related effects, sea surface temperatures	Disease distribution; pathogen development in vector; development, reproduction, activity, distribution and abundance of vectors; transmission patterns and intensity; outbreak occurrence
Leishmaniasis (*Leishmania spp.*)	Sand flies	Temperature, precipitation, El Nino-related effects	Disease incidence and outbreak occurrence; abundance, behavior and distribution of vectors
Chagas disease (*Trypanosoma cruzi*)	Triatomine bugs	Temperature, precipitation, humidity, severe weather event	Vector distribution, increased infestation of houses by vector
Onchocerciasis (*Onchocerca volvulus*)	Black flies	Temperature	Transmission intensity
Arboviruses			
Dengue fever (Dengue virus)	Mosquitoes	Temperature, precipitation	Outbreaks, mosquito breeding ,abundance, transmission intensity (extrinsic incubation period)
Yellow fever (Yellow fever virus)	Mosquitoes	Temperature, precipitation	Outbreaks, incidence; distribution, abundance and breeding of mosquitoes, transmission intensity (extrinsic incubation period)
Chikungunya Fever (Chikungunya virus)	Mosquitoes	Temperature, precipitation	Outbreaks; mosquito breeding and abundance, transmission intensity (extrinsic incubation period)
West Nile virus disease (West Nile virus)	Mosquitoes	Temperature, precipitation	Transmission rates, pathogen development in vector, distribution of disease and vector
Rift Valley Fever (Rift Valley Fever virus)	Mosquitoes	Precipitation, sea surface temperatures	Outbreaks; vector breeding and abundance transmission intensity (extrinsic incubation period)
Ross River virus disease (Ross River Virus)	Mosquitoes	Temperature, precipitation, sea surface temperatures	Outbreaks, vector breeding and abundance, transmission intensity (extrinsic incubation period)
Tick-borne encephalitis (Tick-borne Encephalitis virus)	Ticks	Temperature, precipitation, humidity	Vector distribution, phenology of host-seeking by vector

(continued)

Table 2 (continued)

Disease (Causative Agent)	Vector(s)	Relevant climatic factors	Effects of climatic variability or climate change
Bacterial and Rickettsial Diseases			
Lyme borreliosis (*Borrelia burgdorferi, B. garinii, B. afzelii, or other related Borrelia*)	Ticks	Temperature, precipitation, humidity	Frequency of cases, phenology of host-seeking by vector, vector distribution
Tularemia (*Francisella tularensis*)	Ticks	Temperature, precipitation	Case frequency and onset
Human granulocytic anaplasmosis (*Anaplasma phagocytophilum*)	Ticks	Temperature, precipitation	Vector distribution, phenology of host-seeking by vector
Human monocytic ehrlichiosis (*Ehrlichia chafeensis*)	Ticks	Temperature, precipitation	Phenology of host-seeking by vector
Plague (*Yersinia pestis*)	Fleas	Temperature, precipitation, humidity, El Nino-related events	Development and maintenance of pathogen in vector; survival and reproduction of vectors and hosts; occurrences of historical pandemics and regional outbreaks, distribution of disease

zoonotic diseases, that is, those that spread from animals to humans, is less well studied, although many of the same principles apply to diseases transmitted to humans from both invertebrate and vertebrate animals Mills et al. (2010). Prevailing scientific evidence suggests that there are three primary mechanisms through which climate change can affect vector-borne and zoonotic diseases: (i) geographic range shifts of vectors or reservoirs; (ii) changes in rates of development, survival, and reproduction of vectors, reservoirs, and the pathogens that they carry; and (iii) changes in biting rates of infected vectors or the prevalence of infection in reservoir or vector populations, which affects the likelihood of transmission resulting from contact with a human (Kovats et al. 2001; Mills et al. 2010; Reiter 2001). Several important vector-borne and zoonotic diseases (including their pathogen, etiologic agent, vector, and vertebrate host) that are likely to be affected by climate change are reviewed more extensively below and in (Mills et al. 2010).

3.1.1 Geographic Range Shifts

A study by Rogers and Randolph (2000) modeled the predicted changes in *Plasmodium falciparum* malaria, the most severe form of the disease, under a range of climate scenarios used by the IPCC (HadCM2). Their study was an improvement over previous models, which had based their predictions largely on temperature and rainfall in order to highlight areas suitable for both vector and pathogen development, that is, areas where parasite development occurs rapidly enough to be completed before the vector dies, with bounds defined by habitats suitable for the mosquito. Starting with the present-day distribution of malaria, Rogers and Randolph used a statistical approach to predict disease distribution based on temperature, precipitation, and saturation vapor pressure. Under the "medium–high" scenario, they predicted that 23 million more people would be at risk for malaria by 2050. In contrast, under the "high" scenario (higher mean temperatures), they saw a decrease in exposure of 25 million people. Their findings emphasize the unpredictability of malaria incidence and the possibility of a shift in range rather than an expansion of areas suitable for malaria transmission (Ostfeld 2009).

3.1.2 Vector, Reservoir, and Pathogen Population Dynamics

The populations of vectors, reservoirs, and the pathogens they carry are highly sensitive to climate. There are numerous examples of vector population dynamics in relation to climatic factors. A study of *Ixodes ricinus*, the vector of tick-borne encephalitis and Lyme disease among others, found a 10-fold increase in nymphal activity in spring after unusually high temperatures during the previous summer resulted in faster development of the eggs from that birthing season (Gray 2008). In Brazil, researchers found that the annual incidence of visceral leishmaniasis dips in the year following El Niño and then increases in the second year after El Niño (Franke et al. 2002). They hypothesize that this pattern is due to extended drought conditions during El Niño that lead to a decrease in vector density and subsequent decline in herd immunity followed by a rainy season that triggers an increase in both vector density and infection rate in a high-risk population (Franke et al. 2002). Mosquito populations are affected by precipitation, particularly during breeding, although the impact of changes in rainfall varies by vector species and local landscape characteristics. For example, mosquitoes such as *Anopheles gambiae* that breed in small water pools will be more affected by rainfall than *An. funestus*, which prefer to breed on edges of larger, more stable water bodies (Gage et al. 2008). Another study found that the relationship between precipitation and malaria incidence is positive in upland regions of the Brazilian Amazon, but this relationship reverses along the Amazon river where a 14 cm increase in monthly rainfall can lead to an 80 % decrease in monthly malaria incidence, presumably due to mosquito habitats being washed out by the heavy rains (Olson et al. 2009) (Fig. 7).

Survivorship and reproduction of vertebrate host populations can also be affected by climate-related events. Unusually high temperatures in Australia during the summer of 2002 killed over 3,500 flying foxes (*Pteropus* spp), and at least 18 similar temperature-related die-offs among flying foxes have been recorded in Eastern Australia since 1994 (Welbergen et al. 2008). Long-term hantavirus pulmonary syndrome (HPS) studies in the Southwestern United States have documented consistent patterns between El Niño-related precipitation and the population of the deer mouse reservoir (*Peromyscus maniculatus*) (Glass et al. 2000; Mills et al. 1999; Yates et al. 2002). The "trophic cascade hypothesis" maintains that increased El Niño fall–spring precipitation leads to increases in primary productivity, including vegetation habitat and food resources for deer mice. This increase in resources triggers an increase in deer mouse population density, a direct correlate with increased risk of human infection (Yates et al. 2002).

There are numerous examples of temperature-dependent pathogen development. The extrinsic incubation period of dengue virus—the interval between the acquisition of the infectious agent by its vector and the time when the vector can transmit the infection—was shown to decrease from 12 to 7 days when the incubation temperature of the mosquito vectors was increased from $\leq$30 to 32–35 °C (Watts et al. 1987). Similar results have been found for Western equine encephalomyelitis and St. Louis encephalitis (Reisen et al. 2000), West Nile virus (Reisen et al. 2006), and malaria (Noden et al. 1995).

3.1.3 Pathogen Load of Host and Changes in Transmission Behavior

Temperature and rainfall can affect the pathogen load carried by hosts through a variety of mechanisms. Looking again at HPS, researchers observed an inverse relationship between deer mouse population density and reservoir antibody prevalence when examining temporal data. Successful breeding seasons (spurred by increased precipitation) resulted in populations with a high proportion of

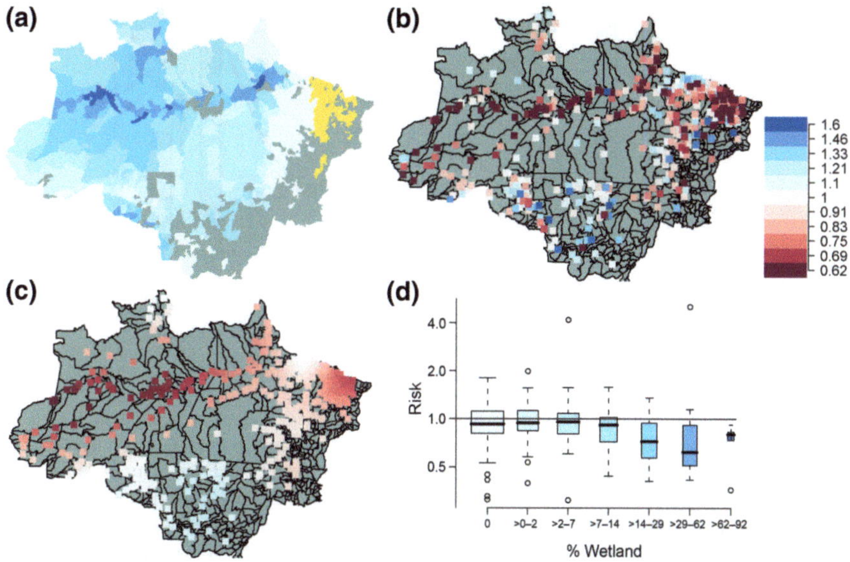

Fig. 7 Precipitation and land cover interactions on malaria risk in the Amazon Basin. Connection of malaria incidence and precipitation risk ratios to wetlands. **a** Percentage of wetlands in Amazon Basin counties (shades of blue), counties without wetlands data (orange), and counties with <80 total malaria cases (gray). Wetland colors correspond to percentage wetland values in panel D. **b** Risk ratios for malaria incidence for 1 SD ($\approx$14 cm) change in monthly precipitation (January 1996–December 1999), plotted at each county seat of government; **c** spatially smoothed risk ratios for $\approx$14-cm changes in monthly precipitation. In both panels, red shaded squares show reduced risk for $\approx$14-cm increase in monthly precipitation; blue shaded squares show increased risk for malaria with increased precipitation. **d** Boxplot of risk ratios for malaria incidence for $\approx$14-cm changes in monthly precipitation, by percentage wetland cover. Box width is proportional to the number of counties in each box. Error bars indicate interquartile ranges, and thick horizontal bars indicate the median. Reproduced with permission from Olson et al. (2009)

uninfected juvenile mice (Mills et al. 1999). Although the high population density led to an increase in the rate of virus transmission, the overall prevalence of the virus in the population remained low because of the continual addition of young, uninfected juvenile mice. However, at the end of the breeding season in late summer, the proportion of young, uninfected mice in the population began to decline as the mice aged and became infected with hantavirus through contact with other mice. The result is that reservoir population density and prevalence of infection are asynchronous. This is referred to as "delayed density-dependent prevalence" (Fig. 8) (Mills et al. 1999). An alternative pathway that links environmental extremes to host infection intensity is through increased stress within a host population. The result can be a decrease in immune response and a higher probably of infection or higher pathogen load within hosts (Mills et al. 2010).

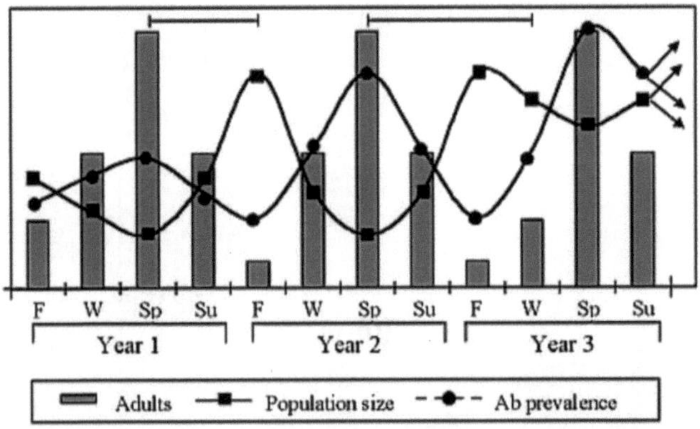

Fig. 8 Explanation seasonal changes in hantavirus prevalence, rodent host population density, and population age structure due to delayed-density-dependent prevalence. A hypothetical schematic of seasonal changes in hantavirus prevalence, rodent host population density and population age structure. In the first autumn, after a normal breeding season, the high density population consists primarily of young not exposed to the virus or recently exposed before development of antibody. Because of deaths in winter, populations decrease to a spring nadir. However, antibody prevalence is high in this population of overwintered adults exposed during the previous breeding season or during winter communal huddling. Unusually favorable conditions during the following spring and summer (first horizontal bar) result in higher population density the second fall, and the increased influx of young of the year (juvenile dilution effect) results in an even lower antibody prevalence than the previous fall. A typical winter results in a high number of winter deaths and a typical low density spring population, but seroprevalence the second spring is higher because of increased opportunities for transmission events in the high density population of the previous fall. The second extended favorable season (second horizontal bar) leads again to high population density and low seroprevalence in autumn. The reservoir population of the third spring demonstrates high antibody prevalence, because of high rates of exposure during the "crowded" conditions of the previous fall, and unusually high population density, because of extended breeding and high overwinter survival. Depending on environmental conditions, the population may abruptly "crash" (if it has exceeded the carrying capacity of the environment) or might continue to increase (e.g., if growing conditions the previous spring and summer resulted in abundant food supplies, such as a mast crop of acorns or pinyon nuts). Reproduced with permission from Mills et al. 1999)

Similarly, there is evidence of viral recrudescence, or reactivation of viral replication, in human viral diseases (Halford et al. 1996; Mehta et al. 2004), likely due to stress-related immunosuppression (Kuenzi et al. 2005; Mehta et al. 2004). Similar viral activity could occur in zoonotic reservoir populations (Mills et al. 2010). Temperature has also been shown to effect arthropod biting rates (Catalá 1991; Patz et al. 1998; Semenza and Menne 2009). Whether through an increased pathogen load in vectors and reservoirs or a change in transmission behavior of vectors, climate-related factors increase the likelihood of pathogen transmission from wildlife to humans for a variety of diseases.

3.2 Diseases of Domestic and Wild Animals and Plants

Many parallels can be drawn between the impact of climate change on the human diseases discussed above and the likely effect of environmental changes on diseases of livestock. Bluetongue virus, a disease transmitted by *Culicoides* midges, has had devastating effects on livestock throughout Africa, the Middle East, Asia, Australia, the United States, and the Mediterranean. In 2006, the disease spread to Central Europe for the first time (Mehlhorn et al. 2007) through an interplay of factors that included an expansion of the range of the primary *Culicodes* vector, temperature dependence of virus replication in the vector (Wittmann et al. 2002), and overlap with a new *Culicodes* species whose range extended 800 km further North than outbreaks had previously been reported (Purse et al. 2005).

The prevalence of parasitical infections of livestock is also subject to environmental changes. The liver fluke trematode (*Fasciola hepatica*) is endemic to Great Britain. (Gale et al. 2009) have identified a confluence of climate-related factors that could affect the prevalence of infection in both sheep and cattle via effects on the parasite, the intermediate host (a snail), and movement of livestock. For example, there is evidence of the liver fluke's ability to adapt to environmental conditions that has enabled them to persist at high altitudes in Bolivia following introduction from Europe (Mas-Coma et al. 2001). Mild winters and high summer rainfall facilitate survival of the intermediate host in regions of Great Britain that have previously been too dry for the snails (Pritchard et al. 2005). Changes in land use and farming practices due to flooding or poor pasture as a result of changing climate may promote transmission of the fluke between cattle and sheep, thereby increasing the frequency of infection transmission among livestock (Gale et al. 2009). There are clear links among livestock health, food safety, and the economic cost of agricultural outbreaks (Paarlberg et al. 2008). These effects can be even more pervasive in regions where animals provide protein, transportation, fuel and clothing, as well as support agricultural labor.

An example of how meteorological information has been used to monitor livestock disease outbreaks is the use of climate and satellite data to predict Rift Valley fever (RVF) in East Africa. RVF is a viral disease that affects domestic animals and humans and is spread by several mosquito species, some of which can transmit the virus directly to their offspring (Linthicum et al. 1999). The relationship between RVF outbreaks and heavy rains that create mosquito breeding sites is well established (Linthicum et al. 1985). Researchers thought that if they could find environmental measurements that are associated with rainfall in East Africa, and that are continually monitored via satellite, they could use these data to forecast RVF outbreaks. They found that when they combined data on equatorial Pacific and Indian Ocean sea-surface temperatures with anomalies in the normalized difference vegetation index (NDVI measures the presence of green vegetation), they would have been able to predict previous RVF outbreaks 1–2 months before viral activity was detected. Continuous environmental monitoring of high-risk areas could trigger livestock vaccination campaigns or insecticidal treatment of known mosquito habitats that could curtail future outbreaks (Linthicum et al. 1999).

Historically, the importance of wildlife diseases has been tied to the threat they pose to human health or livestock (Daszak et al. 2000). Although the diseases cited below are highly unlikely to spill over from wildlife to humans and are less well known than major human pandemics such as cholera, influenza, and smallpox, the implications for human health and well-being are no less significant.

The protozoan parasite, *Ophryocystis elektroscirrha* has been found in North American monarch butterflies and *Danaus gilippus*, the Florida queen butterfly (Altizer et al. 2000). The parasites form dormant spores in the distal third of the adult butterfly abdomen, and heavy parasite loads make it difficult for adult butterflies to open their wings (Leong et al. 1992). As a result, heavily infected adults may die shortly after emergence. Studies have demonstrated the role of migration in lowering the parasite prevalence in butterflies and others have delineated the potential consequences of warming temperature if these migratory species were able to live in one location year-round. (Pascual and Bouma 2009). The role of butterflies as pollinators is widely appreciated but severely understudied (Allen-Wardell et al. 1998). In addition, monarch butterflies make a spectacular, long-distance migration between Mexico and the United States that can involve up to four generations per annual cycle (Brower 1996; Kremen and Ricketts 2000). The regional impacts of a disturbance in this pollinator species are unknown (Allen-Wardell et al. 1998).

Chytridiomycosis is a disease of amphibians that is caused by a fungus, *Batrachochytrium*. The fungus feeds on keratin, grows on amphibian skin, and has been implicated in widespread extinctions (Pounds et al. 2006). The links between the spread of the chytrid fungus and climate change have been hotly debated (Carey and Alexander 2003; Pounds et al. 2006; Rohr et al. 2008). Recent evidence suggests that unpredictable changes in temperatures decrease frog resistance to *B. dendrobatidis* (Raffel et al. 2012). (Kiesecker et al. 2012) argue that pinpointing the causes of these epizootics, or epidemics in animal populations, can be the key to preventing similar outbreaks in human populations. Amphibians are particularly sensitive to environmental changes and thus may be an early warning of outbreaks in human and other animal populations.

Natural resources are an essential "life-support" system for public health, and threats to them affect the long-term sustainability of the human population (McMichael and Beaglehole 2000). For example, massive die-offs of several species of oaks *(Quercus spp.)* have been reported across the Mediterranean region since the early 1980s (Brasier 1996). By 1991, the extent of the decline was revealed in a study of the tree population in Parque Natural de Alcornacales (Cork Oak National Park) in Andalucia, Spain. In the 100,000 ha of oak cover in the park, there were 265 foci where over half the oaks were dead or dying (Brasier 1996). Root excavations of dead trees discovered *Phytophthora cinnamomi* (Brasier 1996), a microscopic soil-borne fungus, and one of the world's most destructive plant pathogens. The fungus requires moisture and warmth to survive, and increased plant stress due to drought may make oaks more susceptible to infection. Indeed, both *P. cinnamomi* infection and drought have been associated with most oak declines in the Mediterranean (Brasier 1996). Projections of *P. cinnamomi* activity and range forecast considerable expansion across Europe with a 1.5 °C increase in annual minimum and maximum temperature (Brasier 1996), a conservative estimate of future warming trends (IPCC 2007). The oak forests of Central Europe are a major renewable timber source (Brasier 1996). They also provide fuel and cork, and are components of traditional agroforestry systems in Spain and Portugal (Brasier 1996).

3.3 Disruption of Synchrony Between Interacting Species

Plants and animals have behavior patterns in response to seasonal environmental suitability; for example, they reproduce when they know that food abundance will peak or migrate when food becomes scarce. The study of these life cycle events in relation to seasonal climate variation is *phenology*. When we reflect on how climate change affects these systems, it is important to realize that the phenology of each trophic level is distinct: insects, plants, and vertebrates each have unique mechanisms that drive their behavior (Visser and Both 2005). However, the activity of one species is often dependent on the behavior of other species in the food chain, and only through natural selection have species evolved a seasonal synchrony that enables them to continue to exist (Visser et al. 2004). When various species respond differently to changes in climate, the result can be a desynchronization, or decoupling, of seasonal activities (Visser et al. 2004).

One of the best-studied phenology systems is that of an oak (*Quercus robur*), the Winter Moth (*Opheroptera brumata*), and the Great Tit (*Parus major*) (Cresswell and Mccleery 2003; Visser et al. 1998). The Great Tit lays its eggs to match the peak abundance of caterpillars that nourish its newly hatched young. In one Great Tit population, the birds did not advance their laying date to match the changes in the caterpillar biomass phenology (Visser et al. 1998), while in another they advanced it too much (Cresswell and Mccleery 2003) (Fig. 9).

Avian migration is another complex phenological process that requires birds to time their migration to match food abundance at their departure location, stopover sites, and overwintering and breeding destinations (Visser and Both 2005). Visser and Both (2005) provide numerous examples of decoupling of migratory bird species movements from the phenology of their main prey, including the pied flycatcher (*Ficedula hypoleuca*), Bewick's swan (*Cygnus columbianus bewickii*), and the American robin (*Turdus migratorius*). American robins are important amplification hosts for West Nile virus (WNV) because they are a highly preferred blood meal source for *Culex* mosquitoes, the vector of the disease Kilpatrick et al. (2006). Robins have been duly named "super spreaders" of WNV in the United States Kilpatrick et al. (2006). Further research showed that a rise in human WNV infections coincided with periods of robin dispersal and migration that led to a sevenfold shift in *Culex* mosquito feeding preferences from birds to humans

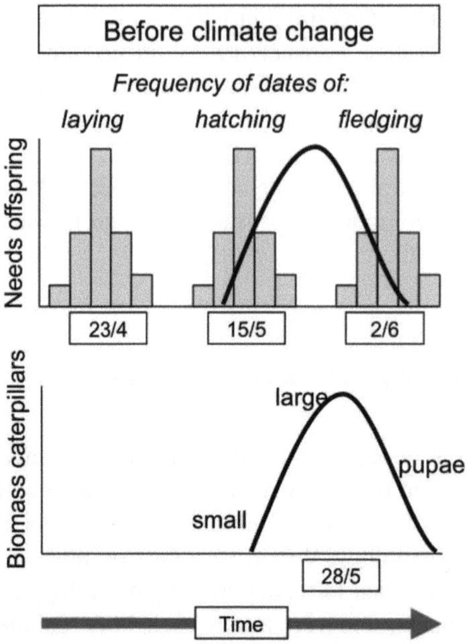

Fig. 9 Climate change induced reproductive mistiming in Dutch Great Tits. A graphical representation of reproductive mistiming due to climate change in Dutch great tits: left panels prior and right panel is during climate change. Top panels represent the frequencies of (from left to right) laying dates hatching dates, and fledging dates. The need for food for the chicks in the nest is indicated with a solid line. Laying dates have not changed under climate change. Lower panels represent the biomass of defoliating caterpillar (main prey for the nestlings) availability: initially low as there are many, but very small, caterpillars, then a peak at the time when there are large caterpillars, followed by a decline when caterpillars start to pupate and are no longer available as prey. The peak date in caterpillar biomass shifts to an earlier date due to climate change, and there is no longer synchronisation between the time the nestlings are fed and maximum food abundance: the population is mistimed. Below the lower panels the environments of decision-making and selection are indicated. Reproduced with permission from Visser et al. (2004)

Kilpatrick et al. (2006). Understanding how environmental changes will affect phenological synchrony of pathogen hosts and vector feeding preferences will be essential information for preventing human zoonotic transmission.

Similar examples of phenological shifts in freshwater systems have been observed where the timing of the diatom phytoplankton bloom has advanced by 27 days over the past four decades, but of the two species of zooplankton that feed on these diatoms, only one has shifted its timing (Winder and Schindler 2004). The result has been a long-term decline in Daphnia zooplankton (the nonshifting species), which may have severe consequences for the aquatic ecosystem (Winder and Schindler 2004). Field studies of plant/pollinator asynchrony have been linked to population crashes and extinctions (Parmesan 2006) with clear implications for human food and natural resources. A scenario that might result from species' differential responses to climate change is the creation of new ecological combinations that could provide opportunities for disease emergence (Gale et al. 2009). Changes in food sources and the presence of species in new places at different times of the year may put wildlife in contact with novel ecological companions that could result in unexpected consequences.

3.4 Trophic Cascades

A trophic cascade defines the concept that links top predators in a food chain to the vegetative biomass at the bottom. When you remove the top predator in a system, a population cascade is triggered: the species in the next trophic level increases, the food source below that species decreases and so on down the food web. In contrast, if you remove the primary producers at the bottom of the food chain, moving up the chain there will be a reduction in the population of each species.

The impact of climate change on the extent of sea ice melt is well documented (IPCC 2007). Data since 1978 show average annual Arctic sea ice area has declined by 2.7 (2.1–3.3) % per decade, with decreases of 7.4 (5.0–9.8) % in summer (IPCC 2007). Atkinson et al. (Atkinson et al. 2004) combined net sampling data on Antarctic krill from 1926 to 2003 to demonstrate the effect of sea ice coverage on krill populations. Controlling for populations of top-down predator and bottom-up resources, they found temporal links between summer krill density and the extent of winter sea ice the year before, perhaps mediated by supporting larval overwintering (Fig. 10). Krill have an enormous role in the entire Arctic ecosystem as one of the primary food sources for penguins, albatrosses, seals, and

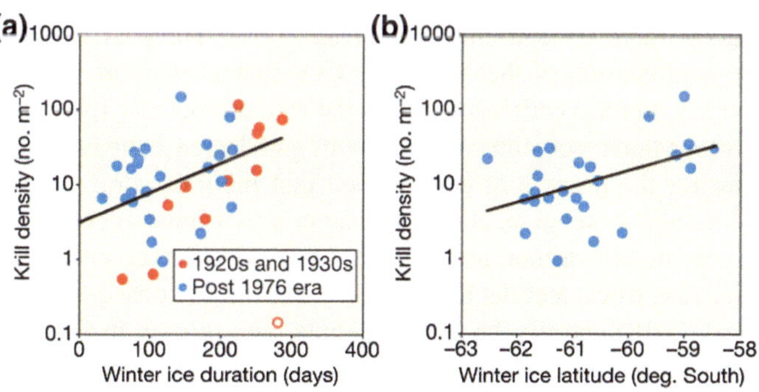

Fig. 10 The relationship between winter sea ice extent and summer krill density. Annual mean density of krill across the SW Atlantic versus **a** sea-ice duration (that is, days of fast ice observed at the South Orkneys the previous winter) and **b** the mean September latitude of 15 % ice cover along a transect across the western Scotia Sea. Regression identified one outlier season (1924, open circle) with exceptionally long ice duration and only 24 net stations, so for the remaining years log10 (no. krill m-2) = 0.49 + 0.0040 (sea-ice duration, days) R2 = 0.21, P = 0.006, n = 35. Log10(no. krill m-2) = 14 + 0.21 (sea ice latitude, degrees) R2 = 0.21, P = 0.02, n = 25. Reproduced with permission from Atkinson et al. (2004)

whales (Atkinson et al. 2004). Consequently, the status of krill is critical for humans who rely on Arctic fauna for their food and livelihood (Fig. 10).

In the Southwestern United States, researchers have found similar climate-induced cascades. Stone et al. (2010) studied the response of the arthropod community to drought stress of the pinyon pine (*Pinus edulis*). They worked in O'Neil Crater, a 55,000-year-old cinder cone with an area less than 1 km^2. Recent drought in the crater killed over half of the pine population and left behind pines on a continuum of health that could be measured by needle retention, the growth rate of the tree, and branch dieback. When they assessed the association between arthropod communities on the pines and the stress level of the trees, they found that both species abundance and richness of the arthropods decreased as the stress level of the pines increased. Over the long term, climate models project a considerable decrease in the range of the pinyon pine, a particularly drought-sensitive species, over the next 80 years (Rehfeldt et al. 2006). Based on their findings in O'Neil Crater, Stone et al. (Stone et al. 2010) suggest that increasing stress levels of the pinyon pine, a common tree across the United States, will likely result in a significant loss of arthropod diversity, with probable cascade effects on species that feed on these insects.

3.5 Alteration or Destruction of Habitat

The majority of emerging infectious diseases are zoonotic, jumping into humans from wildlife reservoirs (Jones et al. 2008), and the role of shrinking wildlife habitat has been implicated as a common causal theme for emerging infections (Daszak et al. 2000). Climate change will have dramatic impacts on many habitats, including coral reefs (Hoegh-Guldberg et al. 2007). The primary threat to these reefs is the acidification of the ocean. As CO_2 increases, more carbon enters the ocean and reacts to form carbonic acid. As the acid dissociates, it forms bicarbonate ions and protons that react with carbonate ions and decrease the carbonate available in the ocean for the process of calcification that rebuilds coral (Hoegh-Guldberg et al. 2007). Fungi (Alker et al. 2001) and bacteria (Kushmaro et al. 1998) that show increased growth at warmer temperatures will also affect coral health as sea temperatures rise. Coral reef decline will have enormous consequences for fisheries (Wilson et al. 2010) and the 2.6 billion people who rely on fish for over 20 % of their protein diet (Brunner et al. 2009).

Warm water and nitrogen favor blooms of marine algae, including two groups, dinoflagellates and diatoms, which can release toxins into the marine environment. These harmful algal blooms (HABs)—also known as red tides—can cause acute paralytic, diarrheic, and amnesic poisoning in humans, as well as extensive die-offs of fish, shellfish, and marine mammals and birds that depend on the marine food web. Over the past three decades, the frequency and global distribution of HABs appear to have increased, and more human intoxication from algal sources has occurred (Van Dolah 2000). For example, during the 1987 El Niño, a bloom of

Gymnodinium breve, previously confined to the Gulf of Mexico, extended Northward after warm Gulf Stream water reached far up the U.S. East Coast, and resulted in human neurologic poisonings from shellfish and in substantial fish kills (Tester et al. 1991). Similarly that year, an outbreak of amnesic shellfish poisoning occurred on Prince Edward Island when warm eddies of the Gulf Stream neared the shore and heavy rains increased nutrient-rich runoff (Hallegraeff 1993).

Modeling in the Netherlands predicts that by the year 2100, a 4 °C increase in summer temperatures in combination with water column stratification would double growth rates of several species of HABs in the North Sea (Peperzak 2005). Biotoxins associated with warmer waters also include ciguatera, which could extend its range to higher latitudes. An association has been found between ciguatera (fish poisoning) and sea surface temperature in some Pacific Islands (Hales et al. 1999).

Some bacteria, especially *Vibrio* species, also proliferate in warm marine waters. Copepods (or zooplankton), which feed on algae, can serve as reservoirs for Vibrio cholerae and other enteric pathogens. For example, in Bangladesh, cholera follows seasonal warming of sea surface temperatures, which can enhance plankton blooms (Colwell 1996). Other *Vibrio* species have expanded in Northern Atlantic waters in association with warm water (Thompson et al. 2004). For example, in 2004 an outbreak of *V. parahaemolyticus* shellfish poisoning was reported from Prince William Sound in Alaska (McLaughlin et al. 2005). This pathogenic species of vibrio had not previously been isolated from Alaskan shellfish due to cold Alaskan waters (McLaughlin et al. 2005). What could have caused the expanded range? Water temperatures during in the 2004 shellfish harvest remained above 15° C and mean water temperatures were significantly higher than the previous 6 years (McLaughlin et al. 2005). Such evidence suggests the potential for warming sea surface temperatures to increase the geographic range of shellfish poisoning and *Vibrio* infections into temperate and even arctic zones.

The observed effect of climate change on terrestrial habitats has also been documented. A recent study in Southwestern Colorado has linked climate-driven forest diebacks in the Southwest with the spread of hantavirus, caused by Sin Nombre virus (SNV) (Lehmer et al. 2012). Sudden aspen decline (SAD) is a phenomenon that has been reported in Arizona (fairweather 2008) and Southern Utah since 2002 (Ohms 2003) and Colorado since 2004 (Worrall et al. 2008) and is characterized by rapid mortality of a mature aspen (*Populus tremuloides*) overstory without subsequent regeneration (Worrall et al. 2010). The primary cause of SAD is thought to be drought events followed by invasion by insects and disease such as the aspen bark beetle (*Trypophleus populi*) and Cytospora cancer (*Valsa sordida*) that attack stressed trees (Worrall et al. 2010).

Lehmer et al. (2012) compared understory plant community structure, small mammal community composition, and SNV prevalence in the small mammal community across the gradient of SAD intensity. They found that in sites with the highest levels of SAD, there was reduced canopy cover, which lead to an increase in understory standing biomass but a decrease in diverse understory microhabitats. When the forest canopy opens up, early successional vegetation that grows well in open/dry conditions forms a dense ground cover that precludes establishment of other types of plants. The diversity in small mammal species was also lowest in high SAD sites, likely owing to the lack of diversity in vegetative cover. A common consequence of habitat disturbance and fragmentation is a shift in small mammal community composition toward dominance by generalist species (Lehmer et al. 2012; Suzán et al. 2009), which is exactly what Lehmer et al. (2012) found in their study sites. The abundance of deer mice (*Peromyscus maniculatus*), the natural SNV reservoir, was highest in sites where aspen death was highest. The prevalence of SNV was also highest in the most disturbed habitats. SNV is spread between deer mice through transfer of bodily fluids, likely as a result of aggressive interactions, such as biting (Calisher et al. 2007). It is likely that the increased density of deer mice may result in higher contact rates among mice, although the time lag for this relationship has not been established (Lehmer et al. 2012).

In summary, the SAD–SNV system shows how a subtle change in temperature and precipitation can lead to successively more severe changes in understory vegetation cover, small mammal community composition, and ultimately to prevalence of a lethal human virus in the wildlife reservoir (Lehmer et al. 2012). This process, referred to as trophic amplification, where interactions across trophic levels can intensify the effects of climate change (Kirby and Beaugrand 2009; Lehmer et al. 2012) highlights the need to understand the interconnected responses of ecosystems, wildlife, and humans to environmental changes in order to prevent harmful impacts on human health.

4 Summary

Understanding the impacts of climate change on human health from a One Health perspective requires working forwards and backwards to connect the species involved in the ecological web linking environmental changes to human health. Climate change has already had a profound effect on biological systems worldwide, and these impacts will be felt by the human population through a variety of mechanisms including: modifications in vector, reservoir, and pathogen lifecycles, and impacts on wildlife and plant diseases, disruptions of synchrony between interacting species, trophic cascades, and alteration or destruction of habitat. Each species responds to environmental changes differently, and in order to predict the movement of disease through ecosystems, we have to rely on expertise from the fields of veterinary, medical, and public health; and these health professionals must take into account the dynamic nature of ecosystems in a changing climate. Rapid environmental changes brought on by climate change intensify the importance of collaborative research and policy-making in order to protect the health of people, animals, and the environment.

References

Alker AP, Smith GW, Kim K (2001) Characterization of *Aspergillus sydowii* (Thom et Church), a fungal pathogen of Caribbean sea fan corals. Hydrobiologia 460:105–111

Allen-Wardell G, Bernhardt P, Bitner R, Burquez A, Buchmann S, Cane J et al (1998) The potential consequences of pollinator declines on the conservation of biodiversity and stability of food crop yields. Conserv Biol 12:8–17

Altizer SM, Oberhauser KS, Brower LP (2000) Associations between host migration and the prevalence of a protozoan parasite in natural populations of adult monarch butterflies. Ecol Entomol 25:125–139

Atkinson A, Siegel V, Pakhomov E, Rothery P (2004) Long-term decline in krill stock and increase in salps within the Southern Ocean. Nature 432:100–103

Battisti DS, Naylor RL (2009) Historical warnings of future food insecurity with unprecedented seasonal heat. Science 323:240–244

Beggs PJ (2004) Impacts of climate change on aeroallergens: past and future. Clin Exp Allergy J Brit Soc Allergy Clin Immunol 34:1507–1513

Bentham G, Langford IH (2001) Environmental temperatures and the incidence of food poisoning in England and Wales. Int J Biometeorol 45:22–26

Bradley NL, Leopold AC, Ross J, Huffaker W (1999) Phenological changes reflect climate change in Wisconsin. PNAS 96:9701–9704

Brasier CM (1996) Phytophthora cinnamomi and oak decline in southern Europe. Environmental constraints including climate change. C.R.L.I.C.A.D.A. Tree Physiol and C.G.M.E.M.N.C.C. Nanceenneeds. Annales des Sciences Forestières 53:347–358

Brower L (1996) Monarch butterfly orientation: missing pieces of a magnificent puzzle. J Exp Biol 199:93–103

Brunner EJ, Jones PJS, Friel S, Bartley M (2009) Fish, human health and marine ecosystem health: policies in collision. Int J Epidemiol 38:93–100

Bumpus HC (1899) The elimination of the unfit as illustrated by the introduced sparrow, Passer domesticus. Biol Lecture Woods Hole Marine Biol Stat 6:209–226

Calisher CH, Wagoner KD, Amman BR, Root JJ, Douglass RJ, Kuenzi AJ et al (2007) Demographic factors associated with prevalence of antibody to Sin Nombre virus in deer mice in the western United States. J Wildl Dis 43:1–11

Carey C, Alexander MA (2003) Climate change and amphibian declines: is there a link? Divers Distrib 9:111–121

Catalá S (1991) The biting rate of *Triatoma infestans* in Argentina. Med Vet Entomol 5:325–333

Checkley W, Epstein LD, Gilman RH, Figueroa D, Cama RI, Patz JA et al (2000) Effect of El Niño and ambient temperature on hospital admissions for diarrhoeal diseases in peruvian children. The Lancet 355:442–450

Colwell RR (1996) Global climate and infectious disease: the cholera paradigm. Science 274:2025–2031

Cresswell W, Mccleery R (2003) How great tits maintain synchronization of their hatch date with food supply in response to long-term variability in temperature. J Anim Ecol 72:356–366

Curriero FC, Patz JA, Rose JB, Lele S (2001) The association between extreme precipitation and waterborne disease outbreaks in the United States, 1948–1994. Am J Public Health 91:1194–1199

Curriero FC, Heiner KS, Samet JM, Zeger SL, Strug L, Patz JA (2002) Temperature and mortality in 11 cities of the eastern United States. Am J Epidemiol 155:80–87

D'Souza RM, Becker NG, Hall G, Moodie KBA (2004) Does ambient temperature affect foodborne disease? Epidemiology 15:86–92

Daszak P, Cunningham AA, Hyatt AD (2000) Emerging threats to infectious diseases wildlife-health biodiversity and infectious. Advancement Sci 287:443–449

Ebi KL, McGregor G (2008) Climate change, tropospheric ozone and particulate matter, and health impacts. Environ Health Perspect 116:1449–1455

Etheridge DM, Steele LP, Francey RJ, Langenfelds RL (1998) Atmospheric methane between 1000 A.D. and present: evidence of anthropogenic emissions and climatic variability. J Geophys Res

Franke CR, Ziller M, Staubach C, Latif M (2002) Impact of the El Niño/southern oscillation on visceral leishmaniasis, Brazil. Emerg Infect Dis 8:914–917

Gage KL, Burkot TR, Eisen RJ, Hayes EB (2008) Climate and vectorborne diseases. Am J Prev Med 35:436–450

Gale P, Drew T, Phipps LP, David G, Wooldridge M (2009) The effect of climate change on the occurrence and prevalence of livestock diseases in Great Britain: a review. J Appl Microbiol 106:1409–1423

Glass GE, Cheek JE, Patz JA, Shields TM, Doyle TJ, Thoroughman DA et al (2000) Using remotely sensed data to identify areas at risk for hantavirus pulmonary syndrome. Emerg Infect Dis 6:238–247

Gray JS (2008) Ixodes ricinus seasonal activity: implications of global warming indicated by revisiting tick and weather data. Int J Med Microbiol 298:19–24

Grinnell J (1917) Field tests of theories concerning distributional control. Am Nat 51:115–128

Gulluk T, Slemr F, Stauffer B (1998) Simultaneous measurements of CO(2), CH(4), and N(2)O in air extracted by sublimation from Antarctica ice cores: confirmation of the data obtained using other extraction techniques. J Geophys Res (Atmos) 103:15971–15978

Hales S, Weinstein P, Woodward A (1999) Ciguatera (fish poisoning), El Niño, and Pacific sea surface temperatures. Ecosyst Health 5:20–25

Halford WP, Gebhardt BM, Carr DJ (1996) Mechanisms of herpes simplex virus type 1 reactivation. J Virol 70:5051–5060

Hallegraeff GM (1993) A review of harmful algal blooms and their apparent global increase. Phycologia 32:79–99

Hoegh-Guldberg O, Mumby PJ, Hooten AJ, Steneck RS, Greenfield P, Gomez E et al (2007) Coral reefs under rapid climate change and ocean acidification. Science 318:1737–1742

Hrudey SE, Payment P, Huck PM, Gillham RW, Hrudey EJ (2003) A fatal waterborne disease epidemic in Walkerton, Ontario: comparison with other waterborne outbreaks in the developed world. Water Sci Technol 47:7–14

International Federation of Red Cross (1998) World disaster report 1997. Oxford University Press, New York

IPCC (2007) IPCC fourth assessment report (AR4)—working group III: climate change 2007. In: Solomon S, Qin D, Manning M, Chen Z, Marquis M, Averyt KB et al (eds) IPCC 4

Jones KE, Patel NG, Levy MA, Storeygard A, Balk D, Gittleman JL et al (2008) Global trends in emerging infectious diseases. Nature 451:990–993

Kiesecker JM, Belden LK, Shea K, Rubbo MJ, Beiden LK (2012) Amphibian decline and emerging disease: what can sick frogs teach us about new and resurgent diseases in human populations and other species of wildlife? Am Sci 92:138–147

Kilpatrick AM, Kramer LD, Jones MJ, Marra PP, Daszak P (2006a) West Nile virus epidemics in North America are driven by shifts in mosquito feeding behavior. R. Ostfelded. PLoS Biol 4:606–610

Kilpatrick M, Daszak P, Jones MJ, Marra PP, Kramer LD (2006b) Host heterogeneity dominates West Nile virus transmission. Proc Roy Soc B Biol Sci 273:2327–2333

Kirby RR, Beaugrand G (2009) Trophic amplification of climate warming. Proc Roy Soc B Biol Sci 276:4095–4103

Kolstad EW, Johansson KA (2011) Uncertainties associated with quantifying climate change impacts on human health: a case study for diarrhea. Environ Health Perspect 119:299–305

Kovats RS, Campbell-Lendrum DH, McMichael AJ, Woodward A, Cox JS (2001) Early effects of climate change: do they include changes in vector-borne disease? Philos Trans R Soc Lond B Biol Sci 356:1057–1068

Kovats RS, Edwards SJ, Hajat S, Armstrong BG, Ebi KL, Menne B (2004) The effect of temperature on food poisoning: a time-series analysis of salmonellosis in ten European countries. Epidemiol Infect 132:443–453

Kremen C, Ricketts T (2000) Conservation issues in international global perspectives on pollination disruptions. Conserv Biol 14:1226–1228
Kuenzi AJ, Douglass RJ, Bond CW, Calisher CH, Mills JN (2005) Long-term dynamics of Sin Nombre viral RNA and antibody in deer mice in Montana. J Wildl Dis 41:473–481
Kushmaro A, Rosenberg E, Fine M, Ben Haim Y, Loya Y (1998) Effect of temperature on bleaching of the coral Oculina patagonica by Vibrio AK-1. Mar Ecol Prog Ser 171:131–137
Lake IR, Gillespie IA, Bentham G, Nichols GL, Lane C, Adak GK et al (2009) A re-evaluation of the impact of temperature and climate change on foodborne illness. Epidemiol Infect 137:1538–1547
Lehmer EM, Korb J, Bombaci S, McLean N, Ghachu J, Hart L et al (2012) The interplay of plant and animal disease in a changing landscape: the role of sudden aspen decline in moderating sin nombre virus prevalence in natural deer mouse populations. EcoHealth 9:205–216
Leong K, Kaya H, Yoshimura M, Frey D (1992) The occurrence and effect of a protozoan parasite, *Ophryocystis elekfroscirrha* (Neogregarinida: Ophryocystidae) on overwintering monarch butterflies, Danaus plexippus (Lepidoptera: Danaidae) from two California winter sites. Ecol Entomol 17:338–342
Linthicum KJ, Davies FG, Kairo A, Bailey CL (1985) Rift Valley fever virus (family Bunyaviridae, genus *Phlebo*virus). Isolations from Diptera collected during an inter-epizootic period in Kenya. The J Hyg 95:197–209
Linthicum KJ, Anyamba A, Tucker CJ, Kelley PW, Myers MF, Peters CJ (1999) Climate and satellite indicators to forecast Rift Valley fever epidemics in Kenya. Science 285:397–400
Lipp EK, Kurz R, Vincent R, Rodriguez-Palacios C, Farrah SR, Rose JB (2001) The effects of seasonal variability and weather on microbial fecal pollution and enteric pathogens in a subtropical estuary. Estuaries 24:266–276
MacKenzie WR, Hoxie NJ, Proctor ME, Gradus MS, Blair KA, Peterson DE et al (1994) A massive outbreak in Milwaukee of cryptosporidium infection transmitted through the public water supply. The N Engl J Med 331:161–167
Mas-Coma S, Funatsu IR, Bargues MD (2001) Fasciola hepatica and lymnaeid snails occurring at very high altitude in South America. Parasitology 123:S115–S127
McLaughlin JB, DePaola A, Bopp CA, Martinek KA, Napolilli NP, Allison CG et al (2005) Outbreak of Vibrio parahaemolyticus gastroenteritis associated with Alaskan oysters. The N Engl J Med 353:1463–1470
McMichael AJ, Beaglehole R (2000) The changing global context of public health. The Lancet 356:495–499
McMichael AJ, Woodruff RE, Hales S (2006) Climate change and human health: present and future risks. The Lancet 367:859–869
Mehlhorn H, Walldorf V, Klimpel S, Jahn B, Jaeger F, Eschweiler J et al (2007) First occurrence of Culicoides obsoletus-transmitted bluetongue virus epidemic in Central Europe. Parasitol Res 101:219–228
Mehta SK, Cohrs RJ, Forghani B, Zerbe G, Gilden DH, Pierson DL (2004) Stress-induced subclinical reactivation of varicella zoster virus in astronauts. J Med Virol 72:174–179
Mills JN, Yates TL, Ksiazek TG, Peters CJ, Childs JE (1999) Long-term studies of hantavirus reservoir populations in the southwestern United States: rationale, potential, and methods. Emerg Infect Dis 5:95–101
Mills JN, Gage KL, Khan AS (2010) Potential influence of climate change on vector-borne and zoonotic diseases: a review and proposed research plan. Environ Health Perspect 118:1507–1514
Noden BH, Kent MD, Beier JC (1995) The impact of variations in temperature on early *Plasmodium falciparum* development in Anopheles stephensi. Parasitology 111:539–545
Ohms S (2003) Restoration of aspen in different stages of mortality in southern Utah. M.S. Thesis. Utah State University, Logan
Olson SH, Gangnon R, Elguero E, Durieux L, Guégan J-F, Foley JA et al (2009) Links between climate, malaria, and wetlands in the Amazon basin. Emerg Infect Dis 15:659–662

Ostfeld RS (2009) Climate change and the distribution and intensity of infectious diseases. Ecology 90:903–905

Paarlberg PL, Hillberg A, Seitzinger, Lee JG, Kenneth H. Mathews J (2008) Economic impacts of foreign animal disease. A report from the economic research service, USDA Number 57

Parmesan C (2006) Evolutionary and ecological responses to recent climate change. Annu Rev Ecol Evol Syst 37:637–669

Pascual M, Bouma MJ (2009) Do rising temperatures matter? Ecology 90:906–912

Patz JA, Strzepek K, Lele S, Hedden M, Greene S, Noden B et al (1998) Predicting key malaria transmission factors, biting and entomological inoculation rates, using modelled soil moisture in Kenya. Tropical Med Int Health 3:818–827

Patz JA, McGeehin MA, Bernard SM, Ebi KL, Epstein PR, Grambsch A et al (2000) The potential health impacts of climate variability and change for the United States. Executive summary of the report of the health sector of the US National Assessment. Environ Health Perspect 108:367–376

Patz JA, Campbell-Lendrum D, Holloway T, Foley JA (2005) Impact of regional climate change on human health. Nature 438:310–317

Patz JA, Vavrus SJ, Uejio CK, McLellan SL (2008) Climate change and waterborne disease risk in the Great Lakes region of the US. Am J Prev Med 35:451–458

Pounds JA, Bustamante MR, Coloma LA, Consuegra JA, Fogden MPL, Foster PN et al (2006) Widespread amphibian extinctions from epidemic disease driven by global warming. Nature 439:161–167

Pritchard GC, Forbes AB, Williams DJL, Salimi-Bejestani MR, Daniel RG (2005) Emergence of fasciolosis in cattle in East Anglia. Vet Rec 157:578–582

Purse BV, Mellor PS, Rogers DJ, Samuel AR, Mertens PPC, Baylis M (2005) Climate change and the recent emergence of bluetongue in Europe. Nat Rev Microbiol 3:171–181

Raffel TR, Romansic JM, Halstead NT, McMahon TA, Venesky MD, Rohr JR (2012) Disease and thermal acclimation in a more variable and unpredictable climate. Nature Clim Change 2:1–6. doi:10.1038/nclimate1659

Rehfeldt GE, Crookston NL, Warwell MV, Evans JS (2006) Empirical analyses of plant-climate relationships for the Western United States. Int J Plant Sci 167:1123–1150

Reisen WK, Meyer RP, Presser SB, Hardy JL (2000) Effect of temperature on the transmission of western equine encephalomyelitis and St. Louis encephalitis viruses by Culex tarsalis (Diptera: Culicidae). J Med Entomol 30:151–160

Reisen WK, Fang Y, Martinez VM (2006) Effects of temperature on the transmission of west nile virus by Culex tarsalis (Diptera: Culicidae). J Med Entomol 43:309–317

Reiter P (2001) Climate change and mosquito-borne disease. Environ Health Perspect 109:141–161

Rogers DJ, Randolph SE (2000) The global spread of malaria in a future, warmer world. Science 289:1763–1766

Rohr JR, Raffel TR, Romansic JM, McCallum H, Hudson PJ (2008) Evaluating the links between climate, disease spread, and amphibian declines. PNAS 105:17436–17441

Schmidhuber J, Tubiello FN (2007) Global food security under climate change. PNAS 104:19703–19708

Semenza JC, Menne B (2009) Climate change and infectious diseases in Europe. Lancet Infect Dis 9:365–375

Solomon S, Qin D, Manning M, Chen Z, Marquis M, Averyt K et al (2007) Climate change 2007—the physical science basis: working group i contribution to the fourth assessment report of the ipcc. Cambridge University Press, Cambridge

Stone AC, Gehring CA, Whitham TG (2010) Drought negatively affects communities on a foundation tree: growth rings predict diversity. Oecologia 164:751–761

Suzán G, Marcé E, Giermakowski JT, Mills JN, Ceballos G, Ostfeld RS et al (2009) Experimental evidence for reduced rodent diversity causing increased hantavirus prevalence. A. Wilbyed. PLoS ONE 4:7

Tester PA, Stumpf RP, Vukovich FM, Fowler PK, Turner JT (1991) An expatriate red tide bloom—transport, distribution, and persistence. Limnol Oceanogr 36:1053–1061
Thomas KM, Charron DF, Waltner-Toews D, Schuster C, Maarouf AR, Holt JD (2006) A role of high impact weather events in waterborne disease outbreaks in Canada, 1975–2001. Int J Environ Health Res 16:167–180
Thompson JR, Randa MA, Marcelino LA, Tomita-Mitchell A, Lim E, Polz MF (2004) Diversity and dynamics of a North Atlantic coastal vibrio community. Appl Environ Microbiol 70:4103–4110
Van Dolah FM (2000) Marine algal toxins: origins, health effects, and their increased occurrence. Environ Health Perspect 108:133–141
Visser ME, Both C (2005) Shifts in phenology due to global climate change: the need for a yardstick. Proc Roy Soc B Biol Sci 272:2561–2569
Visser ME, Noordwijk AJV, Tinbergen JM, Lessells CM (1998) Warmer springs lead to mistimed reproduction in great tits (Parus major). Proc Roy Soc B Biol Sci 265:1867–1870
Visser ME, Both C, Lambrechts MM (2004) Global Climate Change Leads to Mistimed Avian Reproduction. W.F. A Moller and P. Bertholdeds. Adv Ecol Res 35:89–110
Watts DM, Burke DS, Harrison BA, Whitmire RE, Nisalak A (1987) Effect of temperature on the vector efficiency of Aedes aegypti for dengue 2 virus. The Am J Trop Med Hyg 36:143–152
Welbergen JA, Klose SM, Markus N, Eby P (2008) Climate change and the effects of temperature extremes on Australian flying-foxes. Proc Roy Soc B Biol Sci 275:419–425
Wilson SK, Adjeroud M, Bellwood DR, Berumen ML, Booth D, Bozec Y-M et al (2010) Crucial knowledge gaps in current understanding of climate change impacts on coral reef fishes. J Exp Biol 213:894–900
Winder M, Schindler DE (2004) Climate change uncouples trophic interactions in an aquatic ecosystem. Ecology 85:2100–2106
Wittmann EJ, Mello PS, Baylis M (2002) Effect of temperature on the transmission of orbiviruses by the biting midge, Culicoides sonorensis. Med Vet Entomol 16:147–156
Worrall JJ, Egeland L, Eager T, Mask RA, Johnson EW, Kemp PA et al (2008) Rapid mortality of Populus tremuloides in southwestern Colorado, USA. For Ecol Manage 255:686–696
Worrall JJ, Marchetti SB, Egeland L, Mask RA, Eager T, Howell B (2010) Effects and etiology of sudden aspen decline in southwestern Colorado, USA. For Ecol Manage 260:638–648
Yates TL, Mills JN, Parmenter CA, Ksiazek TG, Parmenter RR, Vande Castle JR et al (2002) The ecology and evolutionary history of an emergent disease: hantavirus pulmonary syndrome. Bioscience 52:989
Zhang Y, Bi P, Hiller JE, Sun Y, Ryan P (2007) Climate variations and bacillary dysentery in northern and southern cities of China. The J Infect 55:194–200
Ziska L, Knowlton K, Rogers C, Dalan D, Tierney N, Elder MA et al (2011) Recent warming by latitude associated with increased length of ragweed pollen season in central North America. PNAS 108:4248–4251

Operationalizing One Health: Stone Mountain and Beyond

Carol S. Rubin

Abstract Although the interconnection of humans, animals, and ecosystems has been recognized historically, increasing specialization of professionals in the twentieth century led to decreased communication and collaboration among sectors. In early 2000, a One Health vision of global interconnectedness began gaining in popularity and a series of meetings were held extolling the One Health vision. However, by 2009, detractors were claiming that the One Health approach was indeed all vision and no action. In response to this, international organizations sponsored a carefully planned and structured meeting to construct a way forward that would lead to tangible outcomes. The Stone Mountain meeting, *Operationalizing "One Health": A Policy Perspective—Taking Stock and Shaping an Implementation Roadmap* led to the formation of seven multi-national work groups with defined timelines and outputs. The process has garnered increasing participation and support, and the work groups are on track to demonstrate the value added of a One Health approach.

Contents

C. S. Rubin (✉)
National Center for Emerging and Zoonotic Infectious Diseases,
Centers for Disease Control and Prevention, Atlanta, GA 30333, USA
e-mail: crubin@cdc.gov

Current Topics in Microbiology and Immunology (2013) 366: 173–183
DOI: 10.1007/82_2013_310

Published Online: 5 March 2013

1 Introduction

In the spring of 2010, a conference entitled *Operationalizing "One Health": A Policy Perspective—Taking Stock and Shaping an Implementation Roadmap* was held in Stone Mountain, GA. The international impact of the meeting has been substantial. Whether due to serendipitous timing, careful planning and execution, or the endorsement of essential organizations, the Stone Mountain Meeting (SMM) marked a pivotal point in moving from One Health rhetoric to definitive action steps.

This paper documents the impetus for the meeting, the process that facilitated success and the follow-up activities that the meeting spawned.

2 Progression from Vision to Implementation

In September 2004, a group of strategic thinkers met in New York City and formulated 12 *Manhattan Principles* that called for the international community to adopt a holistic approach to combat "threats to the health of life on earth" under a banner of One World, One Health™ (http://www.hltm.org/docs/HLTM_Twelve_Manhattan_Principles.pdf). These *Principles* identified priorities for leaders and scientists to adopt when faced with the global reality of emerging and re-emerging infectious diseases. Priorities included: recognizing the connections between human, domestic animal, and wildlife health; forming collaborative relationships that foster integration of human and animal surveillance networks; and investing in a mechanism to raise awareness among policy-makers in order to "improve prospects for a healthier planet".

The *Manhattan Principles* provided the vision of a more functional approach to protecting human health during an age of increasing global connectivity. And, indeed, many visionaries accepted the mantle and emerged as spokespersons who articulated the necessity of the One Health approach (http://www.oneworldonehealth.org/sept2004/presentations/eve_foege.html, http://www.oneworldonehealth.org/nov2004/pdfs/newcomb.pdf, http://www.localactionglobalhealth.org/Portals/0/Convergence%20-%20Minnesota%20-%20May%2014%20-%202008%20-%20LONNIE%20KING%20-%20One%20World%20One%20Health%20-%20Presentation.pdf). This was not always an easy sell as most health practitioners functioned within their individual disciplines and were not familiar or comfortable with engaging new colleagues with differing institutional mandates. The strongest advocates originated from animal health sectors, (http://www.avma.org/onehealth/onehealth_final.pdf), although endorsements also came from established human health organizations such as the American Medical Association (http://www.avma.org/onlnews/javma/aug07/070801b.asp) and the American Society of Microbiology (http://asm.org/asm/images/pdf/AtlasPresentation.pdf).

2.1 *Series of Donor Meetings and Progression of Investment in One Health*

In 2007, the Interministerial Conference on Avian and Pandemic Influenza (IMCAPI) was held in New Delhi, India. The importance of cross-sector collaboration as essential to pandemic preparedness was enthusiastically endorsed during that meeting. Given the global anxiety surrounding Highly Pathogenic Avian Influenza (HPAI) H5N1 and the availability of funding for pandemic preparedness, the One Health movement acquired an even larger following, and visionary One Health advocates were met with an expanded audience. Within a short period of time, momentum for a One Health approach was promoted by a series of high-profile national and international meetings that extolled vision but seldom led to action items.

In preparation for IMCAPI 2008 in Sharm-El-Sheikh, Egypt, an international forum was supported by the United Nations Food and Agriculture Orgnization (FAO), the World Organization for Animal Health (OIE), the World Health Organization (WHO), United Nations System Influenza Coordination (UNSIC), United Nations Children's Fund (UNICEF), and the World Bank (WB). This forum developed a *Strategic Framework for Reducing Risks of Infectious Diseases at the Animal-Human-Ecosystems Interface* (http://un-influenza.org/files/OWOH_14Oct08.pdf). The resulting strategy document described the necessity of building upon HPAI H5N1 preparedness to include all emerging infectious diseases, and focused on diseases at the animal, human, and ecosystem interface. This strategic framework was presented at the Sharm-El-Sheikh meeting and a One Health approach was formally endorsed in the meeting summary (http://www.oie.int/doc/ged/D5894.PDF).

2.2 *Winnipeg Meeting: The Tipping Point*

During the Sharm-El-Sheikh meeting, Public Health Agency of Canada's (PHAC), Centre for Food borne, Environmental and Zoonotic Infectious Diseases (CFEZID) offered to host an Expert Consultation to further discuss the strategic framework. The PHAC meeting, *One World One Health™: from ideas to action*, was convened in Manitoba from March 16 to 19, 2009 and brought together almost

200 subject matter experts from 23 countries who shared successes, challenges, and barriers to a One Health approach. Specific recommendations from this meeting included: fostering political will; supporting partnerships and collaboration; encouraging data sharing and integration; building capacity; developing communication strategies and plans; providing incentives for reporting adverse events; encouraging stakeholder and community engagement; and developing supra-country approaches (http://www.phac-aspc.gc.ca/publicat/2009/er-rc/pdf/er-rc-eng.pdf). The meeting summary reads like an action plan, but no tangible follow-up activity was specifically delegated. The bottom-line is that participants went home without assignments, and thus none of the recommendations received coordinated attention.

Three related and impelling events occurred in the months following the Winnipeg conclave. Shortly after the meeting, H1N1 (p2009) emerged globally and reinforced the reality that an influenza virus could surreptitiously jump from animals to humans, and subsequently transverse the globe in a matter of weeks to months. Then, in April 2010, the third IMCAPI meeting was held in Hanoi (http://un-influenza.org/node/4040). The declaration that emerged from this IMCAPI called for a global recognition of the need to "better understand the emergence of disease threats at the animal-human-environment interface through multi-sector actions, and to develop appropriate and sustainable means to reduce such threats". And finally, later in April 2010, FAO, OIE, and WHO released *A Tripartite Concept Note* reinforcing their organization's collaborative intent to combat pathogens at the interface between animal, human, and ecosystem health (http://www.oie.int/fileadmin/Home/eng/Current_Scientific_Issues/docs/pdf/FINAL_CONCEPT_NOTE_Hanoi.pdf).

The succession of high-level meetings and endorsements provided essential defining concepts and objectives; unfortunately, the final outcomes were excellent meeting reports that described actionable steps but did not proscribe responsibility for follow-up. Detractors as well as concerned proponents of the One Health approach voiced the possibility that "One Health" was perhaps just an amorphous concept whose shelf life was expiring.

3 The Stone Mountain Meeting

Recognizing the need for a strategy to move One Health forward toward action steps, leaders from OIE, WHO, and FAO who had participated in the New Delhi, Sharm-El-Sheikh and Winnipeg meetings approached the Centers for Disease Control and Prevention (CDC) requesting that CDC act as the neutral convener of a meeting that would lead to tangible outcomes. The first crucial step was to evaluate previous meetings and consultations, in particular the Winnipeg meeting, to identify why high-level participation, inspiring discussion, and well-written reports had failed to engender action steps.

During discussions with conveners and attendees of the Winnipeg meeting, several impediments to action as well as successful components of the meeting were identified. Impediments to action included the lack of a consensus definition for One Health, the excessive size of the meeting, and the predominance of attendees from the animal health community. Winnipeg, like earlier meetings, invited representatives from many sectors, including public health, food protection, ecosystem health, climate change, and both wildlife and domestic animal representatives. Quite understandably, each of these groups came with their own concept of how One Health should be defined. It is to note that the timing of the Winnipeg meeting coincided with a general belief that in order to be successful, it was necessary that an all-encompassing and uniformly endorsed definition of "One Health" be identified. Many participants related that they felt that, despite what the agenda said, discussion frequently devolved to debating a consensus definition.

The size of the Winnipeg meeting was also reported to be problematic for both planners and attendees.Despite intentions to limit the participation to 60, a compelling draft agenda attracted the attention of highly qualified subject matter experts and the invitee list eventually approached 150 people. This increase in participation also tipped the balance toward a majority of attendees coming from the animal health sector. An imbalance in representation may have led to reiterations of previous discussions and renewals of the existing collaborations rather than formation of innovative alliances. Strengths of the meeting included the use of a strong facilitator who was thoroughly briefed and familiar with the topic (http://conversart.com/) as well as an agenda that allowed sufficient time for working break-out sessions.

After reviewing the lessons learned, the core planning committee decided to move forward with a carefully planned and structured meeting to construct a way forward that would lead to tangible outcomes. The meeting, *Operationalizing "One Health": A Policy Perspective—Taking Stock and Shaping an Implementation Roadmap* was scheduled for May 4–6, 2010 in Stone Mountain, Georgia, USA.

3.1 Structure of the Meeting

The core planning committee pledged to attend weekly conference calls that would adhere to strict agenda items and to distribute assigned action items within 24 h. Decisions included goals and objectives for the upcoming meeting, choice of venue, criteria for identifying participants, as well as defining the agenda and obligations for follow through after the meeting adjourned. Of critical importance, the proposed meeting was anchored to the following overview premise:

> The concept of "One Health" is broad and flexible, as it is intended to encompass the many facets of the relationships among human, animals, and the ecosystems in which they co-exist and interact. In this way, varying detailed interpretations of the scopes of this concept may be put forth according to specific need.

> The discussion at this meeting will focus on defining the actions and policies needed to implement a "One Health" approach. In the context of this meeting, "One Health" represents the inter-sectoral collaborative approach necessary to prevent, detect and control emerging, and re-emerging infectious diseases that exist at the animal-human-ecosystems interface. (cite)

By setting this premise, the planning committee overcame the distracting issue of debating the definition of One Health during the meeting. That discussion was essentially declared to be 'off the table'.

Four meeting objectives were identified:

1. Create a shared view of success for One Health.
2. Take stock of the progress to date in terms of leading practices related to One Health and the policy decisions and financial commitments necessary to support sustainability and expansion.
3. Develop an engagement strategy for key stakeholders to promote One Health.
4. Identify key operational opportunities and barriers to the implementation of "One Health" and develop strategies to address them.

The Evergreen Marriott Conference Resort (http://www.marriott.com/hotels/travel/atleg-evergreen-marriott-conference-resort/) location was chosen, because it provides a remote setting that cloisters participants and encourages ongoing dialog that spills past the meeting agenda. In addition, the dates identified by the planning committee coincided with a limited number of rooms being available. This detail restricted the number of participants to a maximum of 54, thereby overcoming the impediment of inflated participation that was identified during review of the Winnipeg meeting. The attendee list was defined by category of subject matter expertise (i.e., human health, wildlife, domestic animal, economist, and plant health) and geographical representation, but not organizational representation. Definition of invitees by category deliberately conveyed that no single individual was essential to the success of the meeting; the key to moving forward was the balance in areas of expertise created by the attendees. Each organization represented on the core planning committee was assigned a number of attendee slots that it could fill. However, before any invitation was extended, the credentials of that candidate were reviewed, debated, and approved by the entire committee. No invitee was allowed to self-select a replacement. A selection criteria grid mandated that animal, human, and ecosystem health were equitably represented. It is of note that no one on the planning committee knew everyone on the invitee list, and thus it was not simply the same group of people getting together (Fig. 1).

3.2 Outcomes

Early in the SMM, the facilitator challenged participants to focus on short-term goals rather than long-term vision and asked for definition of what success would look like in 3–5 years. The participants agreed that progress toward the One Health vision would be achieved if the following actions were undertaken:

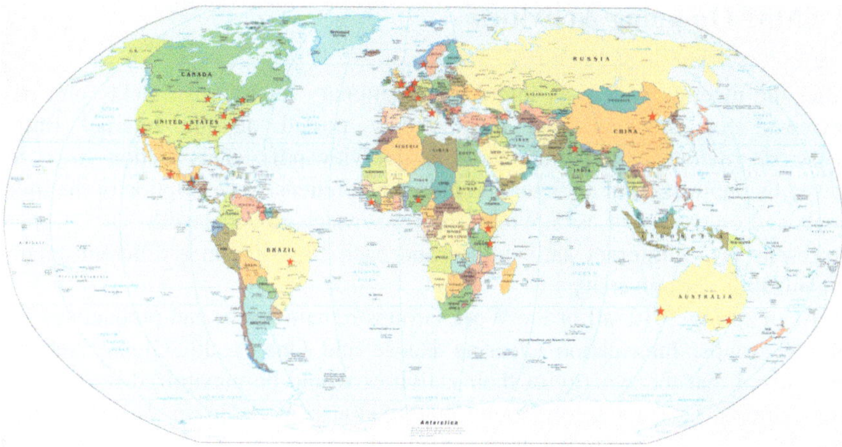

Fig. 1 Map showing geographical distribution of Stone Mountain meeting attendees (note that a single starred location may represent several attendees)

- Initiate culture change manifesting as mutual respect and communication across professions.
- Increase visibility of the One Health approach as adding value.
- Win over political will and funding by demonstrating that One Health can increase impact, especially during periods of finite funding sources.
- Improve coordination and collaboration among sectors for surveillance, outbreak response, and data/sample sharing.

The facilitator further challenged the attendees to identify tangible, results oriented, outcome driven, and practical steps to achieve these short-term goals. The group nominated 21 "enabling initiatives" that would provide positive movement toward the short-term goals and seven of these were selected as most essential: One Health Training; Proof of Concept; Business Plan; Country Level Needs Assessment; Capacity Building; Information Clearing House; and One Health Global Network. Each of these initiatives translated into a Work Group devised to survive the SMM, and near the end of the meeting participants were invited to sign up for membership, and to volunteer for leadership, in a Work Group. Often, at such a juncture, meeting participants politely exit the conference venue. However, at the end of the SMM, every participant signed up to volunteer their time and energy to carry the process forward.

During the final session of the meeting, the Work Groups met to designate co-chairs, draft objectives and deliverables, compose a timeline, and define when the group would next convene. Each Work Group presented this information to the other participants for comment before the SMM adjourned.

4 SMM On-going Activities

After the meeting ended, an initial short summary report was quickly prepared, reviewed for accuracy, and then widely posted on the Internet (http://www.cdc.gov/onehealth/pdf/atlanta/brief_overview.pdf). This action was followed by publication of a longer and more comprehensive description of the initial goals and objectives of each Work Group www.cdc.gov/onehealth. The co-chairs represent seven different nationalities and each comes from a different agency, organization, or university.

As of spring 2012, all of the Work Groups remain active and productive. Two of the groups, Information Clearing House and One Health Global Network, recognized that the synergy of their activities would be maximized if they combined efforts. Thus, the Information Clearing House Work Group was incorporated into the One Health Global Network Work Group. Each Work Group meets independently, primarily by conference call but occasionally in person, and all of the Work Group co-chairs participate in a bi-monthly conference call. CDC facilitates the publication of periodic newsletters summarizing overall Work Group accomplishments, ongoing activities, and specific products from individual Work Groups http://www.cdc.gov/onehealth/pdf/workgroups/newsletter-june-2011.pdf. Selected examples of accomplishments include: the Proof of Concept Work Group conducted an extensive literature review to identify peer-reviewed manuscripts that demonstrate the added value of intervention studies that incorporate animal, human, and environmental health sectors; group summarized their findings in a paper that is currently undergoing clearance. This Work Group has also put out a call for project proposals describing limited scope intervention studies in international settings. The In-Country One Health Self-Assessment Work Group worked with contractors to develop self-assessment guidance document that was reviewed by an expert panel at an April, 2011 workshop; Volume 1 focusing on background and rationale has been completed and Volume 2 focusing on inter-sectoral collaboration is undergoing an additional round of revision. The next step, in collaboration with the Capacity Building Work Group, is to pilot the guidance both in North America (United States and Canada) as well as internationally.

4.1 Expanding Participation

Deliberately limiting attendance at the SMM was deemed essential to achieving its action-oriented goals. However, it was also recognized that the exclusion of accomplished scientists with much to contribute may have inadvertently led to the perception that the SMM process was exclusionary and not representative of the larger One Health community of interest. To foster transparency and garner involvement from a wider array of subject matter experts, the planning committee and the Work Group chairs took every opportunity to present information about the SMM in both formal and informal venues. At the same time, Work Group chairs and Work Group members reached out to colleagues and invited them to join Work Groups, even though they had not attended the seminal SMM. This led to many additional members being added to all of the Work Groups; the Training Work Group expanded from 21 to 52 members.

4.2 Funding to Support Follow-Up

Formation of the Work Groups did not come with dedicated funding nor is there any compensation for the co-chairs; time devoted to this project is in addition to member's ongoing professional obligations. Nonetheless, each Work Group has been creative in its ability to identify funds to cover meetings, consultancies, and reports. For example, The Business Plan Work Group has allied with the University of Georgia Terry College of Business, where Masters of Business Administration students help develop a national One Health business plan. To facilitate the process, the US Department of Agriculture finances an intern to devote additional time to the project. Other Work Groups have sought and received funding from World Bank, US Department of State, FAO, and OIE.

4.3 Coordination with Parallel One Health activities

The SMM was held at a critical juncture that preceded an explosion of complimentary activities originating in other sectors, and all of the Work Groups have prioritized coordination with those other activities rather than any duplication of efforts or competing outputs. For example, the US Agency for International Development Emerging Pandemic Threats (EPT) program RESPOND component is actively working on core competencies and training materials to ensure that future response to outbreaks is coordinated with a One Health approach. At the same time, the University of Minnesota has received funding from the Rockefeller Foundation to look at One Health competencies and curriculum development. Prior to both of these activities, the SMM Training Work Group was developing an online catalog of the existing courses at various institutions. The catalog is being cross-walked with a listing of core competencies. Leadership from these complementary activities are communicating, thus enhancing the value added of the final products.

5 Conclusion: Shifting Paradigms

Some human and animal health practitioners speculate that the current promotion of a One Health approach is simply a return to simpler times when communication and collaboration among disciplines was routine practice that has been inadvertently and perhaps temporarily supplanted by twentieth century specialization (Greaves 2002). Indeed, during the past century there has been an explosion of scientific knowledge that has fostered a separation of human and animal health sectors (Starr 1982). Needless to say, many collaborative relationships persisted but for the most part collaboration was accomplished on a topic-specific basis. The current global movement of people, animals, products, and pathogens demands a holistic approach to surveillance and response to disease emergence and changing ecosystems that is much more than a return to an earlier version of One Health (Cutler 2010; Lloyd-Smith 2009; Feingold 2010).

In his 1962 publication *The Structure of Scientific Revolutions* (Kuhn 1962), Thomas Kuhn used the term paradigm to refer to a set of practices that define a scientific discipline during a specified period of time. He said that a paradigm shift occurs when scientists encounter anomalies which cannot be handled using the prevailing paradigm. He went on to say if there were enough significant anomalies then a crisis would ensue and a new paradigm would need to be generated. The prevailing paradigm for disease control and prevention has not facilitated or promoted coordination among animal, human, and ecosystem sectors. This paradigm was repeatedly tested when the global community was presented with 'anomalies' such as HIV, SARS, and Highly Pathogenic Avian Influenza H5N1. Response to these challenges would have been optimized if there had been coordinated surveillance, response and intervention among human, animal, and environmental health sectors. This lack of overall coordination may have constituted Kuhn's definition of a crisis, and thus provided the basis for consideration of a twenty-first century One Health paradigm. According to Kuhn, the new paradigm is not just a gradual refinement of the old. Rather, it requires deliberate changes in the way scientists approach problems such as emerging pathogens.

It took several years and a series of international meetings to recognize the need and provide the vision for a shift toward a One Health paradigm. The timing is right for moving from One Health rhetoric to definitive action steps and the SMM process and ongoing Work Group activities are critical in this process.

The findings and conclusions in this report are those of the author and do not necessarily represent the official position of the US Centers for Disease Control and Prevention.

Acknowledgment With grateful acknowledgment of the Stone Mountain Meeting Planning Committee:

Drs. Alex Thiermann and Kate Glynn, World Oranization for Animal Health (OIE); Drs. Liz Mumford and Simone Magnino, World Health Organization (WHO); Drs. Juan Lubroth, Jan Slingenbergh and Jim Zingeser, United Nations Food and Agriculture Oganization (FAO); Dr. Alain Vandersmissen, European External Action Service; and Dr. Laura Kahn, Princeton University.

References

Cutler SJ, Fooks AR, van der Poel WH (2010) Public health threat of new, reemerging, and neglected zoonoses in the industrialized world. EID 16:1–7

Feingold BJ, Vegosen L, Davis M et al (2010) A niche for infectious disease in environmental health: rethinking the toxicological paradigm. Envir Health Persp 118:1165–1172

Greaves D (2002) Reflections on a new medical cosmology. J Med Ethics 28:81–85

Kuhn T (1962) The structure of scientific revolutions Chicago. The University of Chicago Press, Chicago

Lloyd-Smith JO, George D, Pepin KM et al (2009) Epidemic dynamics at the human-animal interface. Science 326:1362–1367

Starr P (1982) The social transformation of American medicine. Basic Books Inc, New York

Building a Foundation for 'One Health': An Education Strategy for Enhancing and Sustaining National and Regional Capacity in Endemic and Emerging Zoonotic Disease Management

W. D. Vink, Joanna S. McKenzie, Naomi Cogger, Barry Borman and Petra Muellner

Abstract The rapid global spread of diseases such as SARS, H5N1, and H1N1 influenza has emphasized the pressing need for trans-disciplinary collaboration and cross-border action, and has also exposed a serious deficit of capacity and coordination in dealing effectively with emerging disease threats. The need for capacity development is particularly acute in the developing world, which is the least well-equipped to respond adequately. Such capacity development can be achieved through education and the implementation of applied 'One Health' activities. This chapter describes the establishment of a 'One Health' capacity development program in South Asia, consisting of two phases. The first phase provides Masters level training for public health doctors and veterinarians, with a focus on epidemiology, and disease control. The second phase reinforces the postgraduate training by establishing a sustainable framework for the implementation of collaborative 'One Health' activities such as the development of multi-disciplinary professional networks, implementation of applied zoonotic disease investigation projects, and support for continuing professional development. The objectives are to provide individual skills required to strengthen capacity; to develop an appreciation of the cross-cutting issues which affect human and animal health, set within an institutional context; and to facilitate the development of regional professional networks which will be instrumental in implementing 'One Health' activities.

W. D. Vink (✉) · J. S. McKenzie · N. Cogger · P. Muellner
EpiCentre Institute of Veterinary, Animal and Biomedical Sciences, Massey University, Private Bag 11, 222 Palmerston North, New Zealand
e-mail: W.D.Vink@massey.ac.nz

B. Borman
Centre for Public Health Research, College of Humanities and Social Sciences, Massey University, Private Box 756 Wellington, New Zealand

Current Topics in Microbiology and Immunology (2013) 366: 185–205 185
DOI: 10.1007/82_2012_241

Published Online: 4 July 2012

Contents

1 Introduction

The emergence and spread of major infectious diseases have caused substantial damage to the global economy, while also posing large challenges for public health and disease control authorities (Jones et al. 2008; King et al. 2006; Morens et al. 2004). New infections are more likely to emerge from animal populations than from human populations, and tend to have lower host specificity than established infections. The majority of all emerging human infections are zoonotic (75 % according to Taylor et al. (2001); 60 % according to Jones et al. (2008). There is evidence that the rate of emergence is increasing (Jones et al. 2008): this has been attributed to the escalating interdependency among humans, animals, and the environment, as evidenced by the multiplicity of drivers which have been associated with this trend. These include increased contact rates among human and animal populations, population growth, habitat degradation, climate change, intensification of livestock production, international travel, heightened public awareness, increased animal movements, and trade (Conrad et al. 2009; Osburn et al. 2009; Sherman 2010; Coker et al. 2011; Zinsstag et al. 2011). Furthermore, international trade in live animals and food products contributes to the rapid global spread of zoonotic threats (Fisman and Laupland 2010).

Inevitably, new human and animal pathogens will continue to emerge in the years to come (Woolhouse et al. 2011). The ability to detect and respond adequately to emerging infectious disease threats, in terms of 'on the ground' expertise and capacity as well as human and physical resources, is most constrained in the developing world; paradoxically, this is where the likelihood of such disease emergence is the greatest (Jones et al. 2008). In addition, a significant health issue faced by many developing countries is the persistent high incidence of zoonotic diseases, some of which are considered to be neglected (Anon 2006), and some (such as leptospirosis) are also classified as emerging. Control of these endemic zoonoses has been limited by the lack of integrated application of effective control measures in both animal and human populations (Okello et al. 2011; Sekar et al. 2011).

Resources must be made available to adequately prepare and remain vigilant for the emergence of new diseases (King et al. 2006), and to manage endemic zoo noses (Okello et al. 2011). In the wake of the emergence of HPAI and the pandemic H1N1 influenza strain, there has been a large investment in research, physical resources, and facilities, surveillance, preparedness, and contingency planning for pandemic disease outbreaks. There has also been a large investment in training (Salman 2009); however, the effectiveness of such training for the development of capacity has been constrained by several factors. Training courses tended to be of short duration (days to weeks), which limited the scope and depth that could be covered by such courses. In addition, courses included different participants, and there was considerable variability in the subject matter covered, as well as in the scope, and depth of coverage. Thirdly, there was a lack of assessment of participants' achievement levels, which made it difficult to gage the level of competence attained by the trainees. Opportunities for integrated, coherent, and long-term in-region training remain limited, or insufficiently targeted.

The Centers for Disease Control (CDC) meeting held in May 2010 at Stone Mountain, *Operationalizing 'One Health': A Policy Perspective—Taking Stock and Shaping an Implementation Roadmap*, identified a set of 'critical enabling initiatives', selected training as its top priority (CDC 2010). The 'investment in people' represented by education is an essential complement to the investment in the material requirements for preparing for emerging diseases, and effectuating a response to zoonotic infections. In this context, the CDC has operated the Field Epidemiology Training Program (FETP) since 1975 (Nsubuga et al. 2008; Rolle et al. 2011), and has educated veterinarians through this program; more recently, veterinary FETP programs have been initiated in South-East Asia, with a specific 'One Health' focus (Castellan 2011). This training is directed at field personnel who represent the 'first line of defence'; however, training is required at all levels, including positions related to coordination and decision making (what was described in the CDC Stone Mountain meeting as 'One Health leaders' (Rubin 2011).

The 'One Health' approach aspires to a joint design of disease investigation, control, and management systems for emerging and endemic zoonotic diseases (Kahn 2006), requiring integrated teams of veterinarians, and public health professionals, both in operational, teams and in leadership roles. Effective collaboration involves building new relationships and respect for the roles and expertise of professionals in different sectors (Anon 2008b). It is a truism that 'One Health' action begins with education and collaborative research (Conrad et al. 2009; Osburn et al. 2009), the needs transcend training in one specific discipline or field of specialty, and include cross-cutting issues such as zoonotic diseases, public health, economics, risk assessment and surveillance, and policy development. Furthermore, the development of an enabling environment at institutional, national, and regional levels is equally important for the sustainable implementation of

'One Health' principles. Therefore, capacity development must target multiple levels and work towards the development of sustainable in-country coordinating mechanisms (Anon 2008b).

This chapter describes a two-phase education strategy to strengthen epidemiology and biosecurity capacity in the South Asia region, in a framework that builds the foundations for collaborative 'One Health' action. We discuss experience gained from implementation of the first phase and early stages of the second phase in a regional 'One Health' epidemiology and biosecurity capacity development program in South Asia.

2 A 'One Health' Education Strategy in South Asia

This strategy has been developed to strengthen the epidemiology and biosecurity capacity of animal and human health professionals and institutions that are involved in the management of endemic and emerging zoonoses, and to build the foundations for a 'One Health' approach to zoonotic disease control. A fundamental aspect of this 'One Health' education strategy is the involvement of both public health and animal health professionals in both phases to build collaboration between the two professions.

Phase 1 involves Masters degree training in epidemiology and biosecurity. Objectives of this phase are to: (1) provide relevant training in epidemiology, public health, and biosecurity; (2) build an awareness and understanding of the 'One Health' approach; (3) facilitate effective communication, collaboration, and collegiality among participants with different professional backgrounds within countries, as well as among countries; and (4) strengthen participants' skills in using computers and information and communication technologies.

Phase 2 further develops in-country and regional capacity in epidemiology and biosecurity through regional workshops and by strengthening national institutions that are directly or indirectly responsible for the control of zoonoses. The objectives of Phase 2 are to:

1. Implement collaborative investigations of priority endemic zoonotic diseases in each country involving cross-sectoral teams that effectively operationalize and extend training acquired in Phase 1 to a wider pool of participants, and provide information of value for the adjustment of national disease control policies and enhancement of surveillance activities.
2. Build an operational foundation for multi-sectoral collaboration within and between participating countries and provide a practical context for the provision of international expert assistance and specialized training to further strengthen capacity at both national and regional levels.
3. Demonstrate at both national and regional levels the practical application of a 'One Health' approach that provides information of value to address national and regional 'One Health' priorities.

'One Health' Hubs (OHHs) provide a government-supported organizational and operational framework for implementing Phase 2 of the 'One Health' program by, acting as Centers of Excellence in epidemiology and biosecurity and supporting collaborative networks and investigations of priority zoonoses.

Phase 1 and Phase 2 are integrated to create a transition from the degree of training program into broader operational work that will extend involvement and training in 'One Health' activities to a wider network of professional and scientific personnel in the participating countries and throughout the region. This will broaden capacity and contribute to establishing sustainable 'One Health' activities and build momentum of the 'One Health' agenda in the region.

Massey University is implementing this strategy with the financial support of the Avian and Human Influenza Facility (AHIF) trust fund (Anon 2007), administered by the World Bank. The beneficiary countries are Afghanistan, Bangladesh, Bhutan, India, Nepal, Pakistan and Sri Lanka.

3 The Masters Degree Training Program

3.1 Demographics of the Student Cohort

Phase 1 of the South Asia epidemiology and biosecurity Masters degree program commenced in May 2010, with enrollment of 70 postgraduate students from six of the seven participating countries (Fig. 1). No participants were enrolled from Bhutan due to a shortage of available professional staff at the time. All candidates were identified through the donor's (World Bank's) national programs and Ministerial networks. The candidates were professionals with approved medical and veterinary degrees and relevant experience in disease control activities. Comparable numbers of medical and veterinary professionals were enrolled (34 and 36, respectively). This included 12 women, from India, Nepal and Sri Lanka. Most candidates held posts within the Ministries of Health or Agriculture or governmental research institutes, at mid- to senior level; a number of candidates worked in academic, diagnostic, or clinical capacities. Several of these candidates had previously obtained postgraduate qualifications, or performed epidemiology training. In contrast to training programs which aim to increase field-level capacity, this program targeted professionals working at national and subnational levels.

3.2 Design and Structure of the Degrees

Two Master degrees were specifically established and developed for the purpose of this training: a Master of Public Health (Biosecurity) and a Master of Veterinary Medicine (Biosecurity). A key feature of this training program is that it was

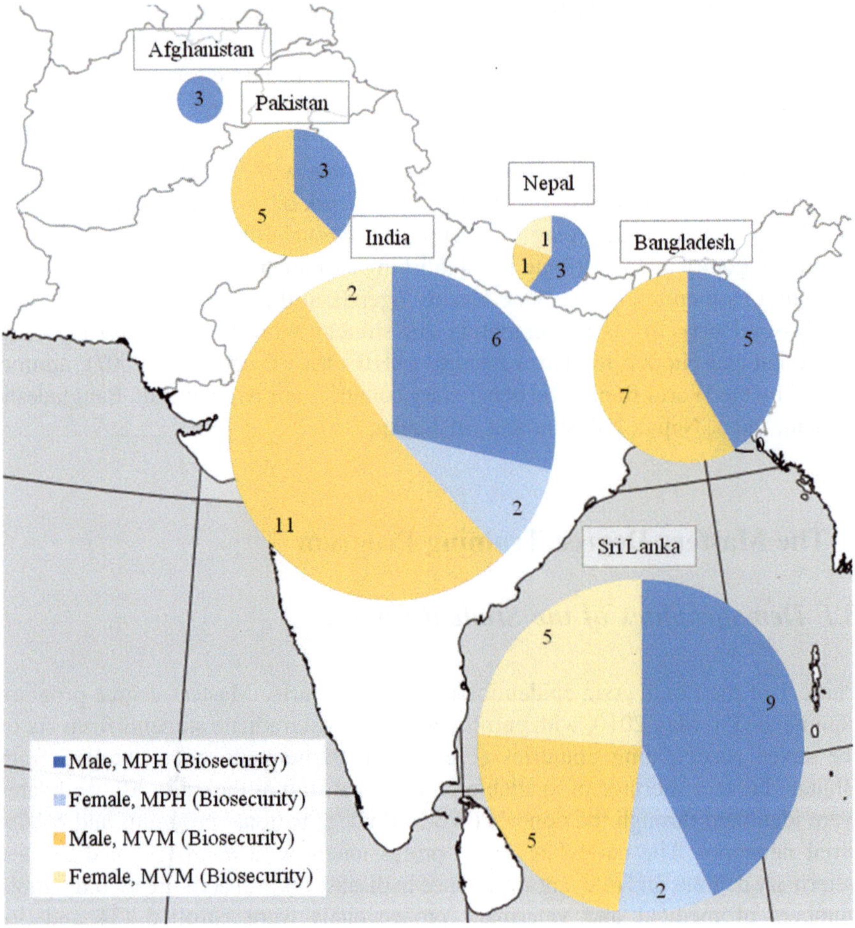

Fig. 1 Student demographic (countries of origin, gender, and degree) of the first cohort of students from the South Asia region ($n = 70$). The sizes of the pie charts are proportional to the numbers of students

specifically designed for delivery by distance methods using Massey University's Internet-based Learning Management System (LMS). This was used to deliver the majority of the course materials and instruction, which allowed candidates to complete the degrees without leaving their day-to-day employment. This was considered essential to avoid further depletion of already scarce in-country professional capability.

Prior to University enrollment for these degrees, the nominated candidates completed a preparatory course to familiarize the prospective students with the LMS and provide essential information on the program. In addition, it was useful as a benchmarking exercise to assess the capabilities of the candidates, particularly in English language competency and computing skills. Subsequent to this course,

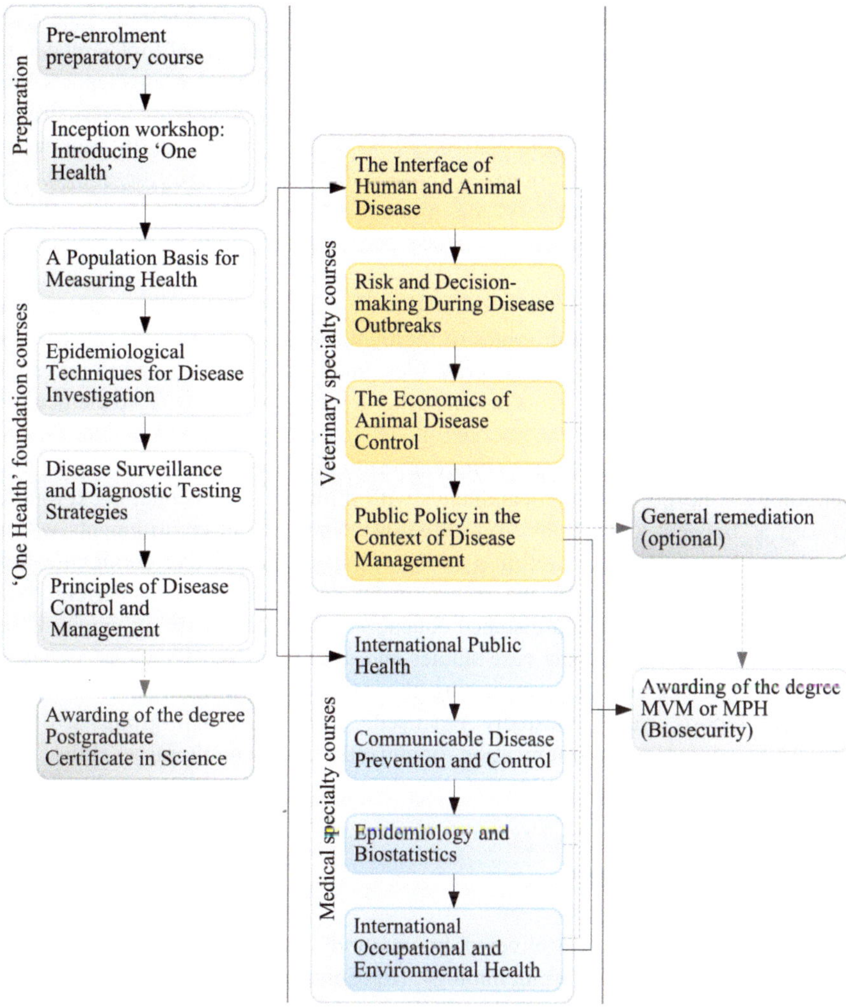

Fig. 2 Program structure of the 'One Health' MVM/MPH (Biosecurity) degrees. All courses were taken sequentially. Double-bordered boxes represent face-to-face events. Candidates completing only the foundation courses were awarded a Postgraduate Certificate in Science; candidates completing all courses were awarded the relevant Master degree. Remediation was given to candidates requiring it

an inception workshop was organized in Singapore. This was an opportunity for the candidates and the program team to meet in person, to present the scope and objectives of the program, and to discuss expectations and requirements.

The degree programs consisted of eight courses which were taken consecutively (see Fig. 2). The courses were run over a 6-week period at an average of 20 study hours per week. The first four courses provided a foundation in epidemiology and public health, and were required to be taken by all candidates in a single learning environment: this was considered critical to achieving the 'One Health' objective. The remaining four courses in each degree addressed specialized topics related to human or animal health to recognize different professional competencies and requirements, and were taken only by the public health doctors or veterinarians, respectively. All courses were delivered online except the fourth foundation course, which had a small online component, but was primarily organized as a face-to-face study conference. As this formed the midpoint of the degree, it also presented a valuable opportunity to assess participants' progress, and prepare for the specialty courses.

The curricula were specifically tailored to be appropriate for the target student demographic; to cover health issues of regional relevance; to include current and cutting-edge scientific knowledge, principles, and understanding; and to make the fullest use of the scientific literature and other learning resources. Two elementary considerations determined the teaching model and methods of instruction. Firstly, the prerequisite of developing a common terminology and understanding of epidemiology concepts (between candidates with different professional backgrounds as well as between candidates from different countries and environments within the South Asia region) resulted in a strong emphasis on effective communication and collaboration. Secondly, the courses were designed following current 'best practice' in pedagogical and distance education techniques, with an emphasis on problem-based learning using case studies and examples. Skills such as searching for, utilizing, and critically appraising the scientific literature and developing technical reports were repeatedly practised. Likewise, small group activities constantly reinforced communication, organization, and collaborative skills.

3.3 Training Delivery

The degree courses were delivered by research groups at Massey University specializing in veterinary epidemiology (EpiCentre) and in public health (the Center for Public Health Research, CPHR) between the period of June 2010 and September 2011. In addition, a number of internationally recognized subject experts made significant contributions to development of the course content and study materials, as well as to the teaching of the courses. The face-to-face component of the fourth foundation course (which was held in New Zealand in December 2010) was structured as a study conference, at which eight international experts presented plenary lectures and subsequently facilitated a related workshop. As the foundation courses were developed and delivered conjointly by veterinary and public health epidemiologists, they sought to give adequate coverage for both MVM and MPH candidates.

All candidates were provided with a laptop to ensure and standardize computer availability. No high-speed broadband connection was required, but access to the Internet was essential due to the predominantly online delivery. The course content was designed to be as bandwidth-efficient as possible, and use of data-intensive applications such as live video, streaming audio, and real-time communications services was avoided.

Although use was made of three standard textbooks, few 'off-the-shelf' print-based resources could be identified that independently gave adequate coverage (in terms of relevance, scope, and applicability) of the course topics. For the 'One Health' foundation courses, this was a consequence of the need for texts that explicitly treated the subject matter in a unified way. The subsequent specialty courses aimed to cover current concepts, approaches and techniques, and in addition to provide content that was of direct relevance for the target audience. For example, the course entitled 'The Interface of Human and Animal Disease' explored pertinent ecological and social factors in human and animal populations that influence the spread of zoonotic disease, in addition to the more obvious factors related to pathogen biology, host pathology and classical epidemiology, or more traditional veterinary public health subjects. Massey University's LMS, which is based on the open-source Moodle platform (Moodle 2012), provided a flexible and adaptable environment which could easily be designed to incorporate content from multiple sources.

The courses were structured around a set of instructional objectives and learning outcomes, which provided a framework that was fleshed out with web-based content synthesized from multiple sources. The courses linked directly into the Massey University library databases and e-journal collection, and the content strategically drew upon the relevant scientific literature, other web-based resources, and the expertise and experience of the candidates themselves. The courses frequently extended beyond strictly technical content to provide candidates with additional skills and knowledge that were considered relevant, for instance drawn from development studies (e.g., logical framework analysis for capacity development) and social science applications (e.g., cognitive mapping, Bayesian belief networks, and multi-criteria decision analysis to explore the concepts of risk and uncertainty). The emphasis of the teaching was very much on application of principles and techniques, and discussion of concepts and relevant case studies. This was supported by the LMS, which provides a rich environment not only for presenting web-based technical content, but for incorporating different activity-based options and for facilitating communication and small-group work. Candidates communicated using tools such as discussion forums, a messaging service, wikis, and real-time chat. Other specialized software and web-based applications were integrated at various points to enhance this environment, such as scenario-based learning software, interactive Flash-based tools and risk analysis software. The courses were intensively supported by teaching staff: each candidate had access to an epidemiologically trained tutor for the duration of each course (with a ratio of six to eight students per tutor), who played an important role in monitoring of the students' activities, as well as for encouraging and facilitating discussion.

As an illustrative example which demonstrates some of the above, a scenario-based simulation exercise was developed within which students were required to play the role of a senior epidemiologist, analyzing data and attempting to determine the cause of an unfolding zoonotic disease outbreak involving human and animal populations. The scenario was based on a real-world event, but details were changed to prevent easy identification. This outbreak was presented in a series of 'episodes' that were developed using specialized e-learning software to present storyboarded scenarios. These 'episodes' were strategically interspersed amongst the teaching material; in each, students were confronted by a sequence of developments and required to react, to revise their hypotheses in response to the evolving patterns in the outbreak and to recommend further investigation and/or control measures. This was supplemented by a combination of group activities including a vote on the putative cause, small-group discussion in forums and the preparation of a report. Full disclosure of the outbreak followed in a face-to-face presentation and discussion during the fourth foundation course. The objectives of this exercise were to present a challenging and immersive learning experience, to provide an authentic insight into the impact of the outbreak on people's lives and livelihoods, and to demonstrate the importance of trans-disciplinary 'One Health' collaboration.

Course assessment was performed using a variety of systems provided by the LMS, and included coursework assignments (usually in essay format), lesson schemas, and quizzes. The assignments, which made up the largest part of the overall assessment, consisted of a combination of individual and group work. Bespoke remediation and additional support was provided to students who required special assistance or were unable to complete activities due to time constraints or exceptional circumstances.

3.4 Outcomes

The attrition rate was low, with 66 of the 70 enrolled candidates receiving degrees. A total of 59 students were awarded Masters degrees (28 MPH and 31 MVM). Seven students who successfully completed the foundation courses but were unable to complete the specialty courses were awarded Postgraduate Certificates in Science. The remaining four students left the program or were unable to complete the foundation courses.

The foundation and specialty courses alike were generally very positively evaluated by the participants. From a total of 229 course evaluation questionnaires submitted by the candidates across all 12 courses, a total of 37candidates rated the courses as being excellent, 43 as very good, 20 as satisfactory, and 0 % as poor. Total 98 % of the candidates felt the learning outcomes had been adequately addressed. Approximately 90 % of the respondents considered the course content, course activities and tutor guidance to be highly relevant or quite relevant, with no differences between the foundation courses and specialty courses. The specialty courses were considered more challenging than the foundation courses, both in terms of the amount and the difficulty of content. This was consistent with the time spent studying, with 56 % of the responses from the specialty courses indicating they spent more than 20 h/week studying versus 29 % for the foundation courses. Participants tended to perform their studies in the evenings and weekends.

Aspects that were particularly appreciated included the collaborative nature of the courses; the activities, which enabled direct application of the material and which represented a different teaching model than most of the candidates were familiar with; and face-to-face interaction during the workshop and study conference. The enthusiasm with which the public health and veterinary cross-communication occurred resulted in a growing spirit of collegiality and receptiveness to collaboration which was truly 'One Health' in nature. This was enhanced by the LMS, which not only facilitated the candidates' increasing competence of the technical aspects, but provided an effective platform for collaboration; it fostered a sense of 'community' and reduced the feeling of isolation that could be experienced by students studying at a distance. A valuable flow-on effect is that this effectively initiated the linkages which will be further developed during the second phase of the program (see below).

Aspects that were less positively evaluated included the restrictions in connectivity and speed of the participants' Internet connections, which affected candidates in several countries and limited the time that could be spent in the LMS. While the participants can continue to access the online courses and materials, a number of participants mentioned the lack of hard copy of study guides, readings and the literature, and other course materials as a limitation. Many people struggled to balance the study commitments with their daily workload. Finally, various elements related to specific content or activities within the courses were negatively evaluated by some individuals.

The general objectives of the training program were first and foremost to provide relevant training in epidemiology, public health, and biosecurity. In the block of foundation courses, the need to reconcile different priorities and focal points of public health and veterinary epidemiology, and make decisions on how to incorporate these into coherent subject matter that had relevance for the enrolled public health professionals as well as veterinarians, gave rise to substantial discussion, and at times exposed differences in perception and approach. The program aims to strike a balance of generic skills required by all, which are taught concurrently to individuals working in public health and in animal health, in a single teaching space, and specific skills required by the two professional disciplines. This required careful consideration.

Developing and delivering the program in such a large and diverse geographical region posed daunting challenges. Notwithstanding technology issues, a mix of languages, cultures and competencies needed to be accommodated. The candidates' English language level was highly variable and at times hampered communication. The diversity of the candidates' professional backgrounds, experience, prior learning levels, and level of seniority, while enriching discussion, influenced their interaction; this was most evident during the online small-group activities, which necessitated a degree of organization and coordination. From the perspective of the development team, a substantial challenge was to calibrate the courses to accommodate the large variability of the candidates' pre-knowledge, such that candidates with little knowledge could be brought up to the minimum level required while simultaneously engaging and extending candidates that already possessed some competence. Other challenges included deciding how much content was required, and related to this, to what level the courses should be taught; which methods of instructionand techniques in the LMS were most effective (how to structure and 'pace' courses, how to present content and design effective activities); and which methods of assessment most accurately assessed the levels of proficiency acquired by the candidates. It was also difficult to develop unified content to illustrate certain principles. For instance, the inclusion of case studies as examples was considered a priority; however, it was frequently difficult to identify public health and veterinary case studies which were equivalent. In many cases, a single case study which best illustrated the matter in hand—whether it pertained to human health or animal health—was used.

4 Applied Training in 'One Health' Hubs

The objective of Phase 2 is to build institutional capacity in epidemiology and biosecurity in the seven beneficiary countries and in the South Asia region. The focus of this phase shifts from teaching individuals within a university learning environment to supporting collaborative 'One Health' activities and applied training of a wider group of animal and public health professionals within a framework that strengthens the government institutions directly or indirectly responsible for diagnosis, preparedness, response, prevention, and control of Highly Pathogenic Avian Influenza (HPAI) and other zoonoses. 'One Health' Hubs (OHHs) provide the organizational and operational framework for this second phase and collaborative investigation projects provide the focus for applied activities and further training supported by an international network of epidemiologists.

Phase 2 includes participants from the Phase 1 training plus an expanded group of new participants from all seven countries. The Phase 1 participants represent a nucleus of professionals who have established relationships with each other, are familiar with communicating and operating in an online environment, have undergone epidemiology training to Masters degree level, understand the benefits of a 'One Health' approach to managing zoonotic diseases, and are motivated to apply these skills to real-world zoonotic disease problems in their countries. A number of these people are in senior management positions or have links to others in these positions, and can take a lead in establishing a government-supported organizational framework and program of activities for Phase 2.

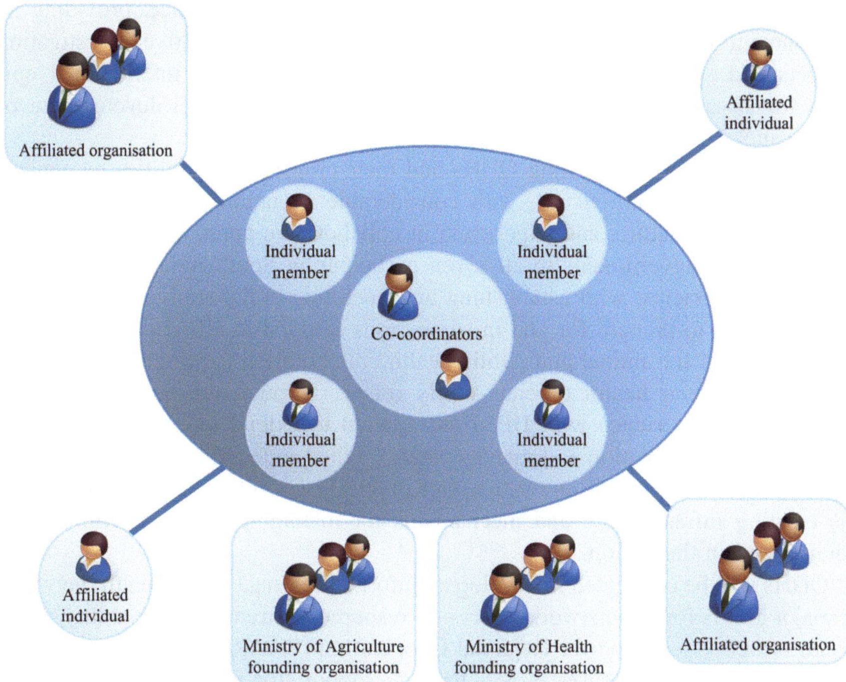

Fig. 3 Organizational structure of 'One Health' Hubs. The Co-coordinators consist of a public health and an animal health professional

4.1 'One Health' Hubs

A single OHH is proposed for each of the seven countries, each OHH comprising a consortium of individuals and organizations that are directly or indirectly involved in the management of zoonotic diseases in human and animal populations. The OHHs are established with the Ministry of Health and the Ministry of Agriculture or Livestock in each country as founding organizations, providing ministerial-level endorsement (see Fig. 3). National and international organizations, including government departments, donor agencies, nongovernmental organizations, research institutes, universities, and zoonoses committees involved in the management and research of zoonoses, are affiliated to the OHHs, facilitating integration with existing national, regional, and international organizations that are involved in the management of endemic and emerging zoonotic diseases. Individual members include medical and veterinary epidemiologists plus other professionals involved in policy, management, delivery and/or research associated with the control of zoonoses. Each OHH will be co-coordinated by an animal health and a public health member. Members will form multi-disciplinary project teams to implement collaborative investigations of priority zoonotic diseases in their country. The OHHs will expand and consolidate regional capacity in public health and epidemiology by functioning as Centers of Excellence, providing an environment for further skills development through collaborative investigation projects and specialized training supported by a network of international epidemiologists and disease experts to assist ongoing professional development of participants.

The process for establishing OHHs and their program of activities recognizes that one size does not fit all, with considerable variation among countries in capacity, infrastructure, and experience in collaborative approaches to managing zoonoses. The governments in all seven participating South Asia countries have had some experience with establishing and operating collaborative organizations and processes to prepare for and in some cases respond to HPAI outbreaks, providing a model for animal and public health collaboration to which senior public health and animal health officials easily relate. Through this program lessons learned from experience with HPAI are applied to extend intersectoral collaboration to the investigation and management of other endemic zoonoses within the OHH in each country. The aim is to establish OHHs that integrate with and support the existing infrastructure and mechanisms for managing zoonoses within each country and in the region.

OHHs will be connected to develop an informal regional 'One Health' network. A major aim is for this network to act as a resource to strengthen the ability of the South Asian Association for Regional Cooperation (SAARC) to manage national and trans-boundary diseases issues, through support for sharing of information, expertise and resources, internally as well as with regional and international organizations such as the Food and Agriculture Organization, and the World Health Organization.

4.2 Collaborative Investigation Projects

Epidemiology and biosecurity skills will be developed beyond the preceding formal education through applied training of a wider group of professionals that is centered around collaborative epidemiological investigations of priority zoonoses in each country. The collaborative investigation projects (CIPs) provide a practical context for extending cross-sectoral collaboration and further building national expertise through the provision of international expert assistance and specialized epidemiology training courses. The aims of the CIPs are to:

- Further, develop epidemiology skills by undertaking field-based investigations;
- Deliver information of value for the adjustment of national disease control policies and enhancement of surveillance activities;
- Provide experience in implementing collaborative investigations and model the benefits of this approach to enhancing the effectiveness of zoonotic disease control programs.

The CIPs are planned in consultation with the governments and other national and regional stakeholders to ensure they focus on zoonoses of national importance and integrate with existing project activities.

Each CIP will be implemented by a multi-disciplinary team with the support of national and international expertise to broaden and strengthen the epidemiology networks in each country and the region. Investigations in the animal and human populations are integrated and designed to complement each other so that the resulting information contributes more to effectively control than nonintegrated epidemiological information from the respective populations. Zoonotic diseases that are being investigated include: rabies (in 3 countries), brucellosis (in 3 countries), leptospirosis, anthrax (in 2 countries), and Crimean Congo Hemorrhagic Fever (CCHF).

Within the 2-year timeframe of the project, participants will be guided in taking the information gained from the field-based investigations and other existing information to recommend enhancements to control policy for the zoonoses of concern. Investigation of the same disease in multiple countries facilitates regional cooperation in investigation and control of these diseases, through sharing of expertise and information. Where possible, participants will be encouraged and supported to publish the results of their investigations and to present these at regional and international meetings and conferences. Applied training courses will be run within the region to enhance specialized epidemiology skills based on the needs identified within countries.

4.3 HubNet

The OHH, CIPs, and extended training are underpinned by HubNet, a web-based platform using state-of-the-art information technologies to provide tools for communication, collaboration, resource-sharing and professional networking, within and among the OHHs. The use of open-source software will enable HubNet to be hosted and maintained in the region, and by the participating OHHs. This will be facilitated by mentoring participants to work with the system to manage their own OHH and CIP sites and other facilities for requesting and sharing information to address disease-related problems that individual participants face. The constraints in Internet access in some countries, with limited bandwidth and intermittent power supply, are explicitly considered in the design of the system.

5 General Discussion

The emergence of pandemic infections such as SARS, HPAI, and H1N1 influenza in the past decade has acted as a catalyst for the increasing acceptance and further development of the 'one medicine' concept, as first conceived by Calvin (Schwabe

1984). The outcomes of 'One Health' activities will comprise of a wide range of outputs related to human health, animal health, food security, nutrition and livelihoods, and environmental sustainability. As a consequence, implementing 'One Health' approaches demands a broad understanding of issues which cut across human and animal medicine, environmental, social, and other sciences, as well as the complexities of their interactions and their impacts on each other. This relies on the willingness and receptiveness of people with a diversity of relevant skills, knowledge and experience to break out of organizational 'information silos' and initiate the dialog out of which the development of such overarching activities can flow. This may motivated by a recognition of the benefits that multi-disciplinary activities can bring (for example, the symbiotic result that can be achieved when technical experts with different specialities contribute towards a single goal), or it may be driven by a managerial imperative to do so, such as the need to utilize resources and implement activities more efficiently through intersectoral and transboundary planning and coordination.

Whichever is the case, motivation and goodwill alone are not sufficient: the capability of professionals with different backgrounds to operate in a multidisciplinary context is dependent on their ability to communicate and collaborateeffectively (Kahn 2011). Inconsistencies or omissions in technical and scientific terminology can make this problematic. The rapid global changes which are transforming the world in which we live have accentuated the interfaces among humans, animals and the environment, and by extension, the growing intersection between human and veterinary epidemiology. Public health and veterinary epidemiologists need to incorporate current approaches and methodologies into a broader vision (Pearce 2009). Communication not only needs to be improved on the scientific and technical level: on the organizational level, the specialized terminology that develops over time within institutional silos (effectively defined as 'jargon'), in combination with the prevailing organizational culture, present formidable impediments to collaboration. Decision- and policy-makers play an essential role in providing leadership, building relationships, trust, and information-sharing mechanisms.

In addition, 'One Health' practitioners must be able to draw on a set of shared competencies . There is currently substantial debate as to what these competencies should be, and how extensive they should be. While it is neither realistic nor desirable to train people who are 'specialists in all disciplines', successful practice requires the acquisition of a nucleus of shared competencies; a set of relevant ability, knowledge, or skills that must extend beyond narrower professional capabilities (Moser 2008). Such competencies can be technical (e.g., consistent application of the principles of epidemiology), or they can be more managerial (e.g., capability for effective leadership, coordination, management, and decision making).

The major international human and animal health agencies, World Health Organization (WHO), Food and Agriculture Organisation (FAO) and World Organization of Animal Health (OIE) have promoted inter-agency and inter-sectoral collaboration through a series of International Ministerial Conferences on Animal and Pandemic Influenza (IMCAPI) which resulted in the drafting of a Tripartite

Concept Note (Anon 2010), the development of a strategic framework for reducing the risks of infectious diseases at the animal-human-ecosystems interface (Anon 2008a), establishing regional collaborative mechanisms and a guide to establishing inter-sectoral collaboration at country levels for information sharing, surveillance, and response (Anon 2008b). Such high-level leadership is necessary to mobilize opinion, solicit political support, and create an impetus. However, it does not create 'One Health' practitioners that are capable of engaging the principles and discussion, integrating these into public health practice at multiple levels, and converting them into meaningful action. In this chapter, we argue that education stands at the basis of 'One Health' capacity development, and is a fundamental requirement for the translation of the concept from theory to practice.

The design of population health programs requires the collaborative input of public health and veterinary professionals (Kahn 2006). There are examples of academic curricula that have incorporated principles of multi-disciplinary health-related teaching, to varying extents, reaching back for decades. In recent years, there has been a sharp increase in the number of academic programs which explicitly aim to do so. However, these are predominantly operated by North American or European institutions (some examples are given by Herrmann and Hershow (2008), Lindenmayer and Schlaff (2008), Conrad et al. (2009), Cribb and Buntain (2009)), and have limited relevance and applicability in a developing country setting, where the shortage of capacity is the greatest, and where the likelihood of emergence of new infections is greatest (Jones et al. 2008). There is an urgent and growing need for trained and capable professionals in the developing countries who will be instrumental in implementing the 'One Health' programs of the future (Asokan et al. 2011). Training is required at all levels, including positions related to coordination and decision making (what was described in the CDC Stone Mountain meeting as 'One Health leaders' (Rubin 2011)).

The establishment and development of the curricula of the Masters degrees described in this chapter was a complex task which demanded intensive collaboration and coordination between personnel within academic groups specializing in public health and veterinary epidemiology, and a number of external subject experts. This was, in itself, an insightful exercise in 'One Health'. The course content was tailored to be relevant for the South Asian participants (e.g., making use of regional case studies where possible); as the degrees operate at a post-graduate level, they build on the existing preknowledge and expertise of the participants. The overarching objective is to create a nucleus of 'One Health'—competent professionals which will make up the initial core of the second phase of the program.

The use of information and communication technologies (ICTs) is a defining feature of the 'One Health' training program. In the first phase, this enabled the participants, many of whom worked in positions of considerable responsibility, to complete the program Masters degree without leaving their duty stations. Although the Internet has become nearly ubiquitous over the past 15 years or so (albeit variable in terms of connectivity, speed and reliability), the penetration and uptake of educational technologies lags relatively behind in the developing world

(Winthrop and Smith 2012). This can be attributed to constraints including limited access to the internet, less familiarity with or receptiveness to the different methods of teaching, and a shortage of human resource capacity (including ICT professionals as well as educational experts). However, the remarkable speed with which internet access has spread globally, and with which social media and online learning technologies have developed, suggests that the 'digital divide' is shrinking (Winthrop and Smith 2012) and makes e-learning feasible in parts of the world in which this was, until quite recently, not the case. Functional and user-friendly open-source software packages such as Moodle (2012) can make distance education vastly more effective, enabling students from developing countries to complete specialized training in-country, thereby avoiding the capacity depletion which occurs when such people are required to leave the country for postgraduate training.

ICTs geared toward education in developing countries have the potential to add value by making positive impacts on a number of outcomes (Wagner et al. 2005), including the acquisition and application of technical skills; the development of computing and technology skills; and other outcomes such as increased innovativeness and increased capability for effective communication. A benefit over 'traditional' distance education is that online systems are specifically designed to provide a collaborative environment, which enhances interaction between teachers and students as well as facilitating exchange of experience and information between students, which contributes toward peer learning and the building of a 'learning community'. This is well-aligned with the 'One Health' approach, which is heavily reliant on effective multi-way communication. Consequently, the degree courses were structured in a way that facilitated interaction, independent application of activities, communication, and active collaboration, among people with different disciplines within a country, among people within the same disciplines but in different countries within the region, and between the participants and leading infectious disease experts across the world. This approach is being continued into the second phase through development of HubNet, a web-based platform that uses open source software to provide tools for communication, collaboration, resource-sharing, and professional networking, within and between the OHHs plus between OHHs and government organizations, international agencies and regional organizations working in related areas.

Research has shown that training programs that focus on application and providing ongoing support, rather than on theoretical knowledge, are more successful (Winthrop and Smith 2012). The Masters degree training was fundamental in laying a strong foundation for the second phase of the program, which provides the opportunity to both consolidate this training and to expand the program to a wider network of animal and human health professionals in each country. The alumnus of the degree program comprises a strongly networked group of human and animal health professionals from varying levels of management within human and animal health institutions throughout the participating countries, who have gained a set of shared competencies in epidemiology and biosecurity, and are capable of promoting interdisciplinary communication and collaboration through a common understanding of definitions, methodologies, and interpretation. The application of these shared competencies to real-world disease problems within the 'One Health' Hub collaborative framework provides the opportunity to create 'One Health' practitioners who have experience and skills in collaborative approaches to the detection, investigation, and management of zoonotic diseases and who are motivated to collaborate by an understanding of the benefits of a 'One Health' approach. As Centers of Excellence in epidemiology, the 'One Health' Hubs support public health, animal health and other professionals to continue developing their skills through sharing knowledge and experience with appropriate expert support and additional training associated with the collaborative investigation projects. Furthermore, these projects provide the opportunity for the more experienced participants to mentor less experienced participants who have an interest in developing their epidemiology and biosecurity skills.

Building sustained and functional collaboration between human and animal health sectors to improve the detection, investigation and management of emerging and endemic zoonoses requires a top-down and a bottom-up approach to provide an enabling environment in which trained 'One Health' practitioners can more effectively share information, knowledge, and resources to manage zoonotic diseases. The network of OHHs established in the second phase of the program described in this chapter provides a mechanism for integrating the collaborative capacity building activities into existing government infrastructures for delivery of human and animal health and for management of zoonoses in each country and in the region. The involvement of key decision makers, policy makers and other stakeholders in establishing and supporting the OHHs and developing the program of activities to ensure it addresses priority zoonoses and critical needs in each country helps build relationships, trust and information-sharing mechanisms between the human and animal health sectors and other international agencies operating in the participating countries. In addition, the OHHs can link into and support field-based epidemiology training programs such as CDC's Field Epidemiology Training Program (FETP) (Nsubuga et al. 2008; Rolle et al. 2011), and FAO's FETPV training for veterinarians (Castellan 2011), and other national and international programs building capacity at the field level. An important objective of Phase 2 is to learn from the experience of applying a 'One Health' approach to the investigation and management of zoonotic diseases, which will be supported by incorporating evaluation of collaborative as well as epidemiological aspects of the investigations.

Acknowledgments This program is financed by the European Commission's Animal and Human Influenza Facility (AHIF), the funding of which is administered by the World Bank. The authors would like to acknowledge the input of the program development team, in particular Ridvan Firestone, Peter Jolly, Cindy Kiro, Marta Mart′nez-Aviles, Roger Morris, Eric Neumann, Eve Pleydell, Lesley Stringer, and Darelle Thomson.

References

Anon (2006) The control of neglected zoonotic diseases. Technical report World Health Organisation, http://whqlibdoc.who.int/publications/2006/9789241594301_eng.pdf Accessed 30 Mar 2012

Anon (2007) Avian and human influenza facility. Technical report Avian and Human Influenza Facility, http://siteresources.worldbank.org/INTTOPAVIFLU/Resources/AHI.Facility.Rocio.May07.pdf Accessed 30 Mar 2012

Anon (2008a) Contributing to one world, one health. A strategic framework for reducing risks of infectious diseases at the animal-human-ecosystems interface. Consultation document produced by FAO, OIE, WHO, UNSIC, UNICEF. Accessed 30 Mar 2012. Technical report, World Bank, http://www.fao.org/docrep/011/aj137e/aj137e00.htm Accessed 30 Mar 2012

Anon (2008b) Zoonotic diseases: A guide to establishing collaboration between animal and human health sectors at the country level. Technical report, World Health Organisation, http://www.searo.who.int/LinkFiles/Publication_Zoonotic.pdf Accessed 30 Mar 2012

Anon (2010) The FAO-OIE-WHO collaboration. A Tripartite concept note. Technical report, FAO, OIE and WHO, http://www.who.int/influenza/resources/documents/tripartite_concept_note_hanoi_042011_en.pdf Accessed 30 Mar 2012

Asokan G, Asokan V, Tharyn P (2011) One health national programme across species on zoonoses: a call to the developing world. Infect Ecol Epidemiol 1:8293.

Castellan D (2011) Development and implementation of multi-disciplinary, multi-sectoral training modules through FETPV. Presentation given at the 1st International One Health Congress, Melbourne, Australia. http://www.onehealth2011.com/pres/Tuesday/Tues%20103%201200%20Castellan%20%5BCompatibility%20Mode%5D.pdf Accessed 30 Mar 2012

CDC (2010) Operationalizing "one health": A policy perspective—taking stock and shaping an Implementation roadmap. Technical Report, Centers for Disease Control, http://www.cdc.gov/onehealth/pdf/atlanta/meeting-overview.pdf

Coker R, Rushton J, Mounier-Jack S, Karimuribo E, Lutumba P, Kambarage D, Pfeiffer DU, Stärk K, Rweyemamu M (2011) Towards a conceptual framework a to support one-health research for policy on emerging zoonoses. Lancet Infect Dis 11(4):326–331.

Conrad PA, Mazet JA, Clifford D, Scott C, Wilkes M (2009) Evolution of a transdisciplinary "One Medicine-One Health" approach to global health education at the University of California, Davis. Prev Vet Med 92(4):268–274.

Cribb A, Buntain B (2009) Innovation in veterinary medical education: The concept of 'One World, One Health' in the curriculum of the Faculty of Veterinary Medicine at the University of Calgary. Rev Sci Tech OIE 28(2):753–762.

Fisman DN, Laupland KB (2010) The 'One Health' paradigm: Time for infectious diseases clinicians to take note? Can J Infect Dis Med 21(3):111–114.

Herrmann JA, Hershow RC (2008) One medicine, one university: The DVM/MPH program at the University of Illinois. J Vet Med Educ 35(2):194–198.

Jones KE, Patel NG, Levy MA, Storeygard A, Balk D, Gittleman JL, Daszak P (2008) Global trends in emerging infectious diseases. Nature 451(7181):990–993.

Kahn L (2006) Confronting zoonoses, linking human and veterinary medicine. Emerg Infect Dis 12(4):556–561.

Kahn LH (2011) The need for one health degree programs. Infect Ecol Epidemiol 1:7919.

King D, Peckham C, Waage J, Brownlie J, Woolhouse M (2006) Infectious diseases: Preparing for the future. Science 313(5792):1392–1393.

Lindenmayer JM, Schlaff AL (2008) The combined Master of Public Health program at Tufts University. J Vet Med Educ 35(2):182–186.

Moodle (2012) Moodle.org: open-source community-based tools for learning. http://moodle.org Accessed 30 Mar 2012

Morens D, Folkers G, Fauci A (2004) The challenge of emerging and re-emerging infectious diseases. Nature 430(6996):242–249.

Moser J (2008) Core Academic Competencies for Master of Public Health Students: One Health Department Practitioners Perspective. Am J Public Health 98(9):1559–1561.

Nsubuga P, White M, Fontaine R, Simone P (2008) Training programmes for field epidemiology. Lancet 371(9613):630–631.

Okello AL, Gibbs EPJ, Vandersmissen A, Welburn SC (2011) One Health and the neglected zoonoses: Turning rhetoric into reality. Vet Rec 169(11):281–285.

Osburn B, Scott C, Gibbs P (2009) One world—one medicine—one health: Emerging veterinary challenges and opportunities. Rev Sci Tech OIE 28(2):481–486.

Pearce N (2009) Research at the interface between human and veterinary epidemiology in occupational and environmental health. In: Proceedings of the 12th symposium of the international society for veterinary epidemiology and economics, Durban, South Africa

Rolle I, Pearson M, Nsubuga P (2011) Fifty-five years of international epidemicassistance investigations conducted by CDC's disease detectives. Am J Epidemiol 174(SUPPL. 11):S97–S112.

Rubin C (2011) Operationalizing one health: the stone mountain meeting. Presentation given at the 1st International One Health Congress, Melbourne, Australia. http://www.cdc.gov/onehealth/pdf/atlanta/australia.pdf Accessed 30 Mar 2012

Salman M (2009) The role of veterinary epidemiology in combating infectious animal diseases on a global scale: The impact of training and outreach programs. Prev Vet Med 92(4):284–287.

Schwabe C (1984) Veterinary Medicine and Human Health, 3rd edn. Baltimore: Williams and Wilkins.

Sekar N, Shah N, Abbas S, Kakkar M (2011) Research options for controlling zoonotic disease in india, 2010–2015. PLoS ONE 6(2):e17, 120

Sherman DM (2010) A global veterinary medical perspective on the concept of one health: focus on livestock. ILAR J 51(3):281–287.

Taylor L, Latham S, Woolhouse M (2001) Risk factors for human disease emergence. Philos Tr R Soc B 356(1411):983–989.

Wagner D, Day B, James T, Kozma R, Miller M, Unwin T (2005) Monitoring and evaluation of ICT in education projects: a handbook for developing countries. Technical report, infoDev/World Bank, http://www.infodev.org/en/Publication.9.html Accessed 30 Mar 2012

Winthrop R, Smith M (2012) A new face of education: bringing technology into the classroom in the developing world. Working paper, Brookings institution, http://www.mobileactive.org/files/file_uploads/01_education_technology_shearer.pdf Accessed 30 Mar 2012

Woolhouse M, Woolhouse ME, Haydon DT, Antia R (2011) How to make predictions about future infectious disease risks. Philos Tr R Soc B 366(1573):2045–54.

Zinsstag J, Schelling E, Waltner-Toews D, Tanner M (2011) From "one medicine" to "one health" and systemic approaches to health and well-being. Prev Vet Med 101:148–156.

Infections at the Animal/Human Interface: Shifting the Paradigm from Emergency Response to Prevention at Source

David L. Heymann and Mathew Dixon

Abstract The majority of emerging infectious diseases have their source in animals, and emergence occurs at the human/animal interface, when infections in animals breech the species barrier to infect humans, the population in which they are often first identified. The response is frequently characterized by a series of emergency activities to contain and manage the infection in human populations, and at the same time to identify the source of the infection in nature. If infection is found to have a source in animals, and if animals cause a continuous threat of human infection, culling is often recommended with severe economic impact. Currently, efforts are being undertaken for closer interaction at the animal/human interface through joint surveillance and risk assessment between the animal and human medicine sectors, and research is underway in geographic areas where emergence at the animal/human interface has occurred in the past. The goal of this research is to identify infectious organisms in tropical and other wild animals, to genetically sequence these organisms, and to attempt to predict which organisms have the potential to emerge in human populations. It may be more cost-effective to learn from past emergence events, and to shift the paradigm from disease surveillance, detection, and response in humans; to prevention of emergence at the source by understanding and mitigating the factors, or determinants, that influence animal infection. These determinants are clearly understood from the study of previous emergence events and include human-induced changes in natural

D. L. Heymann (✉)
Infectious Disease Epidemiology, London School of Hygiene and Tropical Medicine, London, UK
Chatham House Centre on Global Health Security, London, UK
e-mail: David.Heymann@hpa.org.uk

M. Dixon
Chatham House, London, UK

Current Topics in Microbiology and Immunology (2013) 366: 207–215
DOI: 10.1007/82_2012_285

Published Online: 14 December 2012

environments, urban areas, and agricultural systems; raising and processing animal-based foods; and the roles of global trade, migration, and climate change. Better understanding of these factors learned from epidemiological investigation of past and present emergence events, and modeling and study of the cost-effectiveness of interventions that could result in their mitigation, could provide evidence necessary to better address the political and economic barriers to prevention of infections in animals. Such economically convincing arguments for change and mitigation are required because of the basic difference in animal health—driven by the need for profit; and human health—driven by the need to save lives.

Contents

1 Infections at the Animal/Human Interface

Severe Acute Respiratory Syndrome (SARS) was the first major emerging infection identified during the twenty-first century (Parashar and Anderson 2004). A close examination of the outbreak—its origins, the human sickness and death it caused, the national and international responses that occurred, and the effect these responses had on Asian economies—provides a clear lesson of the importance of emerging infections at the animal/human interface, and underscores the reasons that emerging infections must be rapidly detected, assessed, and managed. But understanding and mitigating the factors that align to cause emergence could move this current paradigm of detection, assessment, and response further upstream, to prevention of emergence at its source.

Severe Acute Respiratory Syndrome (SARS) was first detected as a severe atypical pneumonia in the Guangdong Province of China (Heymann and Rodier 2004a). It soon became a burden in hospitals where many patients required respiratory support, and broad-spectrum antibiotics had no effect. As is common with emerging infections, particularly when they present with symptoms common to other known infections, unsuspecting hospital workers became infected. They in turn inadvertently infected family members, and infection then spread to the communities in which they lived (Heymann and Rodier 2004b).

One of these health workers—a medical doctor—travelled to Hong Kong where he stayed in a hotel on the same floor as both Chinese and international guests. Some of these hotel guests became infected. Hypotheses of how they were infected ranged from transmission by aerosols—generated by the infected doctor by cough, sneeze, or vomit—in the corridors or through the hotel ventilation system to shared closed environment such as sharing the same lift (Chan-Yeung and Xu 2003). Some of the infected hotel guests travelled while still in the incubation period, and as illness developed and became serious they were admitted to hospital in Hong Kong, Singapore, Canada, and Vietnam. Hospitalized, they too became the source of infection for hospital workers who in turn served as unintentional amplifiers of transmission to their families and communities.

Molecular and epidemiological investigation suggested that the infection of the index case—never identified—was an onetime event (Walker et al. 2012; Xu et al. 2004). As more information became available, it was further hypothesized that this initial infection was due to close contact with an infected animal, possibly a civet cat, in one of the province's many live (wet) animal markets (Woo et al. 2006). The animal host was thought to have been a carrier of a coronavirus that mutated while replicating, either in the animal or an infected human, in such a way as to cause severe human illness (Wang and Eaton 2007).

The world's interconnectivity through air transport facilitated the international spread of SARS. Precautionary travel advisories were made by the World Health Organization (WHO) recommending that people avoid unnecessary travel to countries where outbreaks were occurring, and by July 2003, just over 7 months after the SARS coronavirus was thought to have emerged, human to human transmission had been interrupted and the outbreak was declared over (Heymann 2006).

SARS resulted in 8,422 probable infections and 916 (11 %) deaths (Chan-Yeung and Xu 2003). The economic impact of the outbreak on GDP was estimated at US$30–100 billion from decreased commerce, travel, and tourism (Keogh-Brown and Smith 2008). Unlike HIV, which is thought to have emerged during the late nineteenth or early twentieth century, the SARS coronavirus did not become endemic, and economic recovery was rapid.

SARS and other emerging infections share a common theme: infection is often first detected in human populations in which an emergency containment response occurs, most times before the source of infection is understood. Initial recommendations for control are thus based on what evidence is available from the current outbreak or previous outbreaks caused by similar organisms. They are of necessity precautionary, and often severe. And as for SARS, the burden and response can cause a wide-ranging negative impact to economies.

If it were possible to identify infectious agents carried by wild and domestic animals and to predict if, when and where they would emerge in humans, and if these animals could then be somehow removed from contact with humans or cleared of infection, human sickness and death could be prevented and economies protected. Studies are underway to identify and characterize infectious organisms in wild animals in geographic sites where emerging infections are known to have occurred in the past (Grace et al. 2012; Jones et al. 2008; UC Davis: Vet Medicine 2009). Though it is possible through these studies to understand the variety of infectious agents carried by wild animals, prediction of which organisms will emerge in human populations using genetic sequence or other information will likely be very challenging and as yet is not possible (Biek and Real 2010).

Moving further upstream, investigation of individual emergence events can identify the risk factors, or determinants, that align to cause the putative breaches in the animal/human species barrier. If these risk factors could in some way be mitigated, the risk of future emergence could be decreased. The current paradigm of emergency response, and the concurrent attempts at prediction and prevention, could then be shifted further upstream.

2 Shifting the Paradigm

In the case of SARS, there was a flurry of field research activity in the Guangdong Province during and just after the outbreak, but over time funding decreased and research slowed. Among the research that was completed was a study of workers in some of the province's wet markets that suggested that up to 22 % (12/55) had antibody evidence of a coronavirus infection related to the SARS coronavirus, but that none had a history of severe respiratory symptoms such as were occurring in persons with SARS (Parry 2003). Further field research might have helped to better understand the risk factors for emergence, but it was not conducted, and the epidemiology remains unclear.

Risk factors for emergence, in addition to being a market worker as suggested by the completed study, might also include being a hunter of wild animals, being a restaurant worker who kills and or butchers/prepares wild animal meat for consumption, or being a member of a household who buys live or recently killed wild game meat from a wet market (Weiss and McMichael 2004; Wolfe et al. 2007).

Even though evidence is available from just one epidemiological study of SARS, a series of actions outside the human and animal health sectors could be useful in preventing a future outbreak in the Guangdong Province from another emerging pathogen (Daszak et al. 2012; Wood et al. 2012). These include education of all those who come into contact with wild game (and domestic animals) about how to protect themselves against infection; regulation with enforcement of wet markets and eating establishments that does not drive these activities underground, but rather ensures safe animal handling; and regulation and enforcement of trade between hunters and markets, and between markets and those who purchase. Other activities might be research to determine whether wild animals (e.g. civet cats) could be raised commercially under conditions that prevent their infection and risk to humans—or further downstream, more effective education of health workers about infection control. This latter activity would ensure that if other actions such as those above fail to prevent emergence, amplification of transmission of emergent organisms could be prevented.

Risk factors for emergence events caused by a more broad range of organisms might occur in sectors such as plant agriculture, community planning, water, and sanitation. Human migratory dynamics, land-use approaches, and the influence of climate and manipulation of natural ecosystems can also amplify known risks, and create novel emergence pathways (Patz et al. 2008).

Mitigation of the risk factors for emergence thus requires a focussed and collaborative effort across multiple disciplines—a one health approach, as defined by the American Veterinary Medical Association (2008) (American Veterinary Medical Association 2012). Emergence may occur among humans living and working in small rural farming communities carved out of tropical rain forests, savannah, mountains, and desert that are in close proximity to wild animals, or to domestic animals they tend that have been in close proximity to wild animals. Outbreaks of Nipah and Ebola Reston Virus infection in pigs raised in unprotected environments in Malaysia and the Philippines, respectively, are an example, and both outbreaks spilled over into human populations (Luby et al. 2009; Miranda and Miranda 2011).

Emergence may occur in larger urban communities where human contact with animals is limited to a few farm animals in close proximity to households, to domestic pets, or to rodents and other animals that have adapted to the urban environment (Alirol et al. 2011). Animals come in contact with humans or other animals as they range (e.g. cows and chickens in parts of Asia) or browse (e.g. urban foxes and rodents) (Bradley and Altizer 2007). The continued high rate of contact between humans and poultry in both smaller backyard farms and larger market system farms continues to permit repeated human exposure to the H5N1 influenza A virus that is endemic in poultry stock. Children and adults are thought to have been infected by contact with living chickens in backyards, and adults have been shown to become infected at some point during the process of raising or slaughtering/butchering chickens (Kerkhove et al. 2011).

Risk factors of emergence in these settings are lacking or inadequate community planning, lack of understanding by populations about risks associated with animal contact, failure to adopt and adhere to safe farming and slaughter/food processing and preparation practices, and failure to maintain sanitation and water infrastructure. Mitigation across all these sectors would require empowering communities to develop a safer living environment through urban planning, developing and maintaining robust water and sanitation infrastructure, controlling rodent and other animal populations in both peri-urban and urban areas, ensuring safe animal husbandry, and providing understanding of risks through community-based education (Fobil et al. 2012).

Risk factors for emergence also occur all along the food chain. Growing demand for animal-based food has led to the ever more complex food chains that involve live animal processing and trade networks (Schlundt et al. 2004). Prevention of infectious disease emergence through the food chain and agricultural system requires understanding of the risks at each step along the pathway from the farm to the fork. If infectious agents pass through the food chain and enter foods, their impact can be minimized at the final intervention point, where animal-derived foods can be prepared carefully in the factory, restaurant, and household either by cooking or other means to remove or mitigate the risk of infection. Others must be controlled earlier—during the period animals are being raised, during slaughter, and during transport (Collins and Wall 2004).

Climate change also appears to be a factor in emergence of human infection. Rainfall associated with ENSO (El Niño/*Southern Oscillation*) in East Africa, for example, has contributed to frequent outbreaks of Rift Valley fever as a result of flooding that increases breeding sites of the mosquito vector (Anyamba et al. 2009). The frequency of leptospira transmission from rodents to humans has been shown to increase during events in Latin America, Bangladesh, and India following heavy rains and flooding (Lau et al. 2010). Lassa fever has also emerged after severe drought in Sierra Leone, when rodents carrying the Lassa virus were forced to move closer to humans so that they could survive on agricultural products in cultivated fields or storage facilities, contaminating human food supplies (Bonner et al. 2007).

Risk factors related to climate change are multiple, and in addition to more robust civil engineering projects to prevent flooding and channel water for irrigation, better rodent and wild animal control is required, as is continued participation in the negotiation of the International Climate Control Treaty (Tol et al. 2007).

Finally, overuse of antibiotics in livestock animals is thought to be a risk factor for the emergence of antimicrobial-resistant bacteria in animals. Though there is still much debate within the scientific community as to the contribution of antibiotics in farming systems to the rise of antibiotic resistance, the implications on emerging antibiotic resistance in human populations is even less well understood (Barton 2000). But there is general consensus that farming systems are likely to contribute to the flow of antibiotic residues and resistant microbes in the wider ecosystem and in humans by runoff into water used or consumed by humans, especially in economically poor settings where farming communities exist alongside densely populated human environments with poor sanitation/sewage systems (Abraham 2011; Segura et al. 2009). Clearly, cross-sector action is required to mitigate these risks using the example of the connection between antimicrobial resistance in both animal and human sectors provides a key lesson for ensuring interdisciplinary planning is incorporated when designing zoonotic control strategies.

3 The Opportunity

There is an opportunity to learn from past emergence events, and from those that are presently occurring or will occur in the future. Application of what has been learned can help shift the paradigm from detection, assessment, and response to prevention at the source. But solid evidence must be available or obtained, assessed for risk, and used. There is a great amount of scientific knowledge about the risk factors of emergence and their mitigation already available from previous investigation and risk assessment. Much more can be obtained from in-depth study

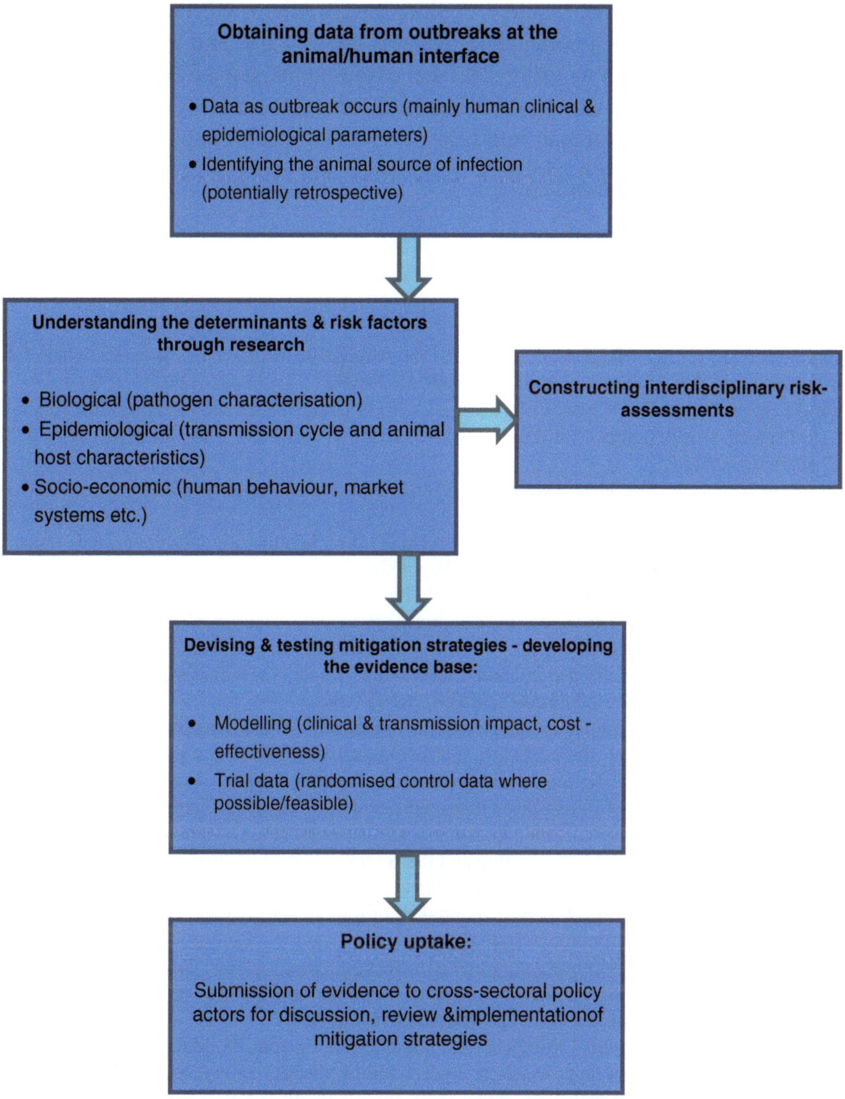

Fig. 1 Transforming evidence at the animal/human interface into policy, a simple flow chart

of each emergence event as it occurs. Research must also take into account human behavior, and ensure that populations most at risk clearly understand the measures required to reduce or protect behavior that is high risk. Many emergence events occur in well-defined geographical areas involving the poorest communities, so designing interventions and strategies that are cost-effective and sustainable will be imperative.

Many of the measures required to shift the paradigm will encounter political barriers, especially when commercial benefits are at stake, and these barriers will need to be broken down by using clear and easy to understand evidence from cost-effectiveness and of a variety of risk mitigation strategies (Fig. 1). By working together at the animal/human interface using a one health approach, emergence events in the future can be decreased, and lives and economies saved.

References

Abraham W-R (2011) Megacities as sources for pathogenic bacteria in rivers and their fate downstream. Int J Microbiol. doi:10.1155/2011/798292

Alirol E, Getaz L, Stoll B, Chappuis F, Loutan L (2011) Urbanisation and infectious diseases in a globalised world. Lancet Inf Diseases 11(2):131–141 (Elsevier Ltd.) doi:10.1016/S1473-3099(10)70223-1

American Veterinary Medical Association (2008) One health: a new professional imperative, pp 1–76

American Veterinary Medical Association (2012) One health—it's all connected. https://www.avma.org/KB/Resources/Reference/Pages/One-Health.aspx

Anyamba A, Chretien J-P, Small J, Tucker CJ, Formenty PB, Richardson JH, Britch SC et al (2009) Prediction of a rift valley fever outbreak. Proc Natl Acad Sci U S A 106(3):955–959. doi:10.1073/pnas.0806490106

Barton MD (2000) Antibiotic use in animal feed and its impact on human health. Nutr Res Rev 13(2):279–299. doi:10.1079/095442200108729106

Biek R, Real LA (2010) The landscape genetics of infectious disease emergence and spread. Mol Ecol 19(17):3515–3531. doi:10.1111/j.1365-294X.2010.04679.x

Bonner PC, Schmidt W-P, Belmain S R, Oshin B, Baglole D, Borchert M (2007) Poor housing quality increases risk of rodent infestation and Lassa fever in refugee camps of Sierra Leone. Am J Trop Med Hyg 77(1):169–175. http://www.ncbi.nlm.nih.gov/pubmed/17620650

Bradley Ca, Altizer S (2007) Urbanization and the ecology of wildlife diseases. Trends Ecol Evol 22(2):95–102. doi:10.1016/j.tree.2006.11.001

Chan-Yeung M, Xu R-H (2003) SARS: epidemiology. Respirology 8 Suppl S9–14. http://www.ncbi.nlm.nih.gov/pubmed/15018127

Collins JD, Wall PG (2004) Food safety and animal production systems: controlling zoonoses at farm level. Revue scientifique et technique (International Office of Epizootics) 23(2):685–700. http://www.ncbi.nlm.nih.gov/pubmed/15702728

Daszak P, Zambrana-Torrelio C, Bogich TL, Fernandez M, Epstein JH, Murray KA, Hamilton H (2012) Interdisciplinary approaches to understanding disease emergence: the past, present, and future drivers of Nipah virus emergence. Proc Natl Acad Sci U S A 1–8. doi:10.1073/pnas.1201243109

Fobil JN, Levers C, Lakes T, Loag W, Kraemer A, May J (2012) Mapping urban malaria and diarrhea mortality in accra, ghana: evidence of vulnerabilities and implications for urban health policy. J Urban Health Bull New York Acad Med. doi:10.1007/s11524-012-9702-x

Grace D, Mutua F, Ochungo P, Kruska R, Jones K, Brierly L, Al E (2012) Zoonoses project 4 mapping of poverty and likely zoonoses hotspots. http://mahider.ilri.org/bitstream/handle/10568/21161/ZooMap_July2012_final.pdf?sequence=4

Heymann DL (2006) SARS and emerging infectious diseases: a challenge to place global solidarity above national sovereignty. Ann Acad Med Singapore 35(5):350–353. http://www.ncbi.nlm.nih.gov/pubmed/16830003

Heymann DL, Rodier G (2004a) Global surveillance, national surveillance, and SARS. Emerg Inf Dis 10(2):173–175. doi:10.3201/eid1002.031038
Heymann DL, Rodier G (2004b) SARS: a global reponse to an international threat. Brown J World Aff X(2):185–197
Jones KE, Patel NG, Levy MA, Storeygard A, Balk D, Gittleman JL, Daszak P (2008) Global trends in emerging infectious diseases. Nature 451(7181):990–993. doi:10.1038/nature06536
Keogh-Brown MR, Smith RD (2008) The economic impact of SARS: how does the reality match the predictions? Health Policy (Amsterdam, Netherlands) 88(1):110–120. doi:10.1016/j.healthpol.2008.03.003
Kerkhove MDV, Mumford E, Mounts AW, Bresee J, Ly S, Bridges CB, Otte J (2011) Highly pathogenic avian influenza (H5N1): pathways of exposure at the animal-human interface, a systematic review. Methods 6(1):1–8. doi:10.1371/journal.pone.0014582
Lau CL, Smythe LD, Craig SB, Weinstein P (2010) Climate change, flooding, urbanisation and leptospirosis: fuelling the fire? Trans Royal Soc Trop Med Hyg 104(10):631–638. doi:10.1016/j.trstmh.2010.07.002
Luby SP, Gurley ES, Hossain MJ (2009) Transmission of human infection with Nipah virus. Clin Inf Dis Off Publ Inf Dis Soc Am 49(11):1743–1748. doi:10.1086/647951
UC Davis: Vet Medicine (2009) USAID: Predict. http://www.vetmed.ucdavis.edu/ohi/predict/index.cfm. Accessed 22 Oct 2012
Miranda MEG, Miranda NLJ (2011) Reston ebolavirus in humans and animals in the Philippines: a review. J Inf Dis 204 Suppl (Suppl 3):S757–760. doi:10.1093/infdis/jir296
Parashar UD, Anderson LJ (2004) Severe acute respiratory syndrome: review and lessons of the 2003 outbreak. Int J Epidemiol 33(4):628–634. doi:10.1093/ije/dyh198
Parry J (2003) Asymptomatic animal traders prove positive for SARS virus. BMJ (Clinical research ed.) 327(7415):582
Patz JA, Olson SH, Uejio CK, Gibbs HK (2008) Disease emergence from global climate and land use change. Med Clin North America 92(6):1473–1491 xii. doi:10.1016/j.mcna.2008.07.007
Schlundt J, Toyofuku H, Jansen J, Herbst SA (2004) Emerging food-borne zoonoses. Revue scientifique et technique (International Office of Epizootics) 23(2):513–533. http://www.ncbi.nlm.nih.gov/pubmed/15702717
Segura PA, François M, Gagnon C, Sauvé S (2009) Review of the occurrence of anti-infectives in contaminated wastewaters and natural and drinking waters. Environ Health Perspect 117(5):675–684. doi:10.1289/ehp.11776
Tol RSJ, Ebi KL, Yohe GW (2007) Infectious disease, development, and climate change: a scenario analysis. Env Dev Econ 12(05):687–706. doi:10.1017/S1355770X07003841
Walker P, Cauchemez S, Hartemink N, Tiensin T, Ghani AC (2012) Outbreaks of H5N1 in poultry in Thailand: the relative role of poultry production types in sustaining transmission and the impact of active surveillance in control. J R Soc Interface doi:10.1098/rsif.2012.0022
Wang LF, Eaton BT (2007) Bats, civets and the emergence of SARS. Curr Top Microbiol Immunol 315:325–344. http://www.ncbi.nlm.nih.gov/pubmed/17848070
Weiss RA, McMichael AJ (2004) Social and environmental risk factors in the emergence of infectious diseases. Nat Med 10(12 Suppl):S70–S76. doi:10.1038/nm1150
Wolfe ND, Dunavan CP, Diamond J (2007) Origins of major human infectious diseases. Nature 447(7142):279–283. doi:10.1038/nature05775
Woo PC, Lau SK, Yuen K-Y (2006) Infectious diseases emerging from Chinese wet-markets: zoonotic origins of severe respiratory viral infections. Curr Opin Inf Dis 19(5):401–407. doi:10.1097/01.qco.0000244043.08264.fc
Wood JLN, Leach M, Waldman L, Macgregor H, Fooks AR, Jones KE, Restif O et al (2012) A framework for the study of zoonotic disease emergence and its drivers: spillover of bat pathogens as a case study. Philos Trans Royal Soc London Ser B Biol Sci 367(1604):2881–2892. doi:10.1098/rstb.2012.0228
Xu R-H, He J-F, Evans MR, Peng G-W, Field HE, Yu D-W, Lee C-K et al (2004) Epidemiologic clues to SARS origin in China. Emerg Inf Dis 10(6):1030–1037. doi:10.3201/eid1006.030852

One Health from a Social–Ecological Systems Perspective: Enriching Social and Cultural Dimensions

Helen Ross

Abstract This chapter offers insights from the environmental management paradigm of 'social–ecological systems' and related bodies of theory on people–environment relationships to assist the evolution of the One Health interdisciplinary endeavour of health promotion across human–animal ecosystem relationships. It also seeks to expand thinking about the social and cultural dimensions that are likely to prove important in the development of thinking and practice in the One Health field. It advocates consideration of cultural and economic relationships affecting people's interactions with domesticated and wild animal species and ecosystems, and exploration of the cognitive and behavioural aspects of these interactions.

Contents

H. Ross (✉)
The University of Queensland, St Lucia, Brisbane, QLD, Australia
e-mail: Helen.Ross@uq.edu.au

Current Topics in Microbiology and Immunology (2013) 366: 217–229
DOI: 10.1007/82_2012_275

Published Online: 3 November 2012

1 Introduction

This chapter offers insights from the environmental management paradigm of 'social–ecological systems' to assist the evolution of the One Health interdisciplinary endeavour of health promotion across human–animal ecosystem relationships. It then seeks to expand thinking about the social and cultural dimensions that are likely to prove important in the development of thinking and practice in the One Health field, especially tracing the nature of interactions among humans, domestic and wild animal species, and their ecosystems.

The interdisciplinary One Health agenda and goals for system strengthening (FAO et al. 2008; APEC 2011) imply the need for a systems approach and antecedent disciplines, particularly epidemiology, which rely heavily on systems understandings in tracing the sources of disease outbreaks and identifying risk factors that are important in their prevention. This chapter suggests ways in which One Health's systems approaches can learn from many tenets of the paradigm of social–ecological systems which is gaining currency in environmental management (Berkes and Folke 1998; Gunderson et al. 1995; Gunderson and Holling 2002; Walker et al. 2004; Walker and Salt 2006; Westley et al. 2002; Berkes et al. 2003), while also identifying ways in which One Health can stretch the thinking of social–ecological systems scholars (Cumming 2010; Dudley 2008). Recognising current limitations in the social–ecological systems paradigm, the chapter also draws on other bodies of literature that deal with human–environment relationships, bringing in important social science concepts that the social–ecological systems literature has yet to incorporate well.

A second, and equally important purpose is to elaborate social and cultural dimensions of people–animal–environment relationships that One Health theorists and practitioners are likely to find important. There is a need both for individual studies that focus on relevant social dimensions (e.g. communication) and for comprehensive frameworks or sets of concepts to look towards ensuring strategic approaches to understanding and practising in One Health.

2 Social–Ecological Systems: Interpreting Complex Adaptive Systems at Multiple Scales

The social–ecological systems paradigm (Walker and Salt 2006) provides a particularly useful foundation for the evolution of One Health (Cumming 2010), for its adoption of the 'complexity' paradigm and focus on key concepts such as 'resilience' and 'adaptive capacity'. As others have observed, the dynamic infection behaviour of avian influenza (HPAI) suggests complex adaptive systems rather than predictable linear behaviour (Cumming 2010; Dudley 2008). The idea of social–ecological systems derives from ecologists, observing forest behaviour. Since it is ecosystem based, the theory also tends to be highly 'place-based', and thus requires translation for circumstances such as pandemics, where diseases that may erupt in one place quickly transfer to others, and indeed for species behaviour across diverse locations, as with migratory birds (Caron et al. 2010). Thus, in One Health the social–ecological system of interest is both local and global, with disease manifestations—and opportunities to prevent or address challenges—occurring at local to regional and national scales, quickly crossing scales and jumping geographical boundaries.

The key tenets of social–ecological systems thinking are that:

(1) Social–ecological systems are complex and adaptive systems; they do not behave in predictable linear ways (Walker and Salt 2006).
(2) The social–ecological aspects represent coupled parts of a single system (Berkes and Folke 1998; Folke 2006), i.e. they are inseparable and equally important.
(3) These systems are nested, with patterns at any one 'level' (scale) affecting and being affected by others, particularly the adjacent levels (for instance, the relationships among household, community and region).
(4) There is no stability or preferred state. Rather, a system (at any level, and across levels) can tend towards one or more states (e.g. ecosystem conditions in which a pathogen is barely present), yet be prone to 'flip' into other states—desirable or otherwise—particularly if thresholds (such as in a seasonal change or temperature rise that allows that pathogen to thrive) are approached (see Si et al. 2010 for a study of seasonal vegetation change, wild bird movements and presence of avian influenza H5N1).
(5) System behaviour is made complex by the interactions of fast and slow variables. For instance, social and cultural change may be slow, but economic crises or disease outbreaks can happen fast.
(6) We should be interested in the resilience of these systems, and their transformations into other patterns (some scholars separate the idea of transformation from resilience, in the sense of persistence). How can desirable transformations occur, and devastating ones be avoided? We should be interested both in 'specified resilience' (e.g. resilience of a system to a particular condition, such as 2° climate change), and in 'generalised resilience' to multiple possible circumstances (Walker and Salt 2006; Walker and Westley 2011).
(7) No single authority can govern such systems. 'Adaptive governance' and 'adaptive co-management', generally involving multiple collaborating parties, is necessary, and this adjusts to requirements as the system changes (Olsson et al. 2006, 2007).
(8) Social learning is highly important in the management of social–ecological systems (Wilson 2012).

The theorising in social–ecological systems is still at early stages. Social scientists are pressing for greater elaboration of the social dimensions (Davidson 2010), recognising the neglect of vital concepts such as cognitions (Jones et al. 2011), power relationships (Berkes and Ross 2013) and to spell out the dynamics of social change (Cote and Nightingale 2012). Thus, it is useful to draw on other interdisciplinary paradigms which deal with people–environment (though seldom animal) relationships (see Table 1). Human ecology, closely allied with anthropology and ecology, contributes strong understandings of how cultures have co-evolved with particular types of ecosystem, and focuses on some key ecological concepts such as adaptation. Political ecology, stemming from human ecology and more particularly political economy, brings in the notion of power relationships in the distribution of access to natural resources. Environment-behaviour studies (environmental psychology), combine psychology, architecture and geography in a strongly transactional view of how people affect and transform environments (through their cognitions, behaviour patterns and the making of physical changes), and how those environments affect people by shaping their potential for action (Ross et al. 2000). The work under this paradigm is particularly relevant to the built environment, and to understanding the rationales behind people's behaviour patterns.

2.1 Opportunities for One Health

Given the origins in forest ecology, the social–ecological systems concept is strongly place based. One Health requires the nested idea of local to global relationships (including individuals), but cannot afford to limit to particular local and regional ecosystems. Rather, One Health will be interested in changes in social–ecological conditions that favour zoonotic disease transfers and spread (e.g. Caron et al. 2010; Si et al. 2010), and in pandemics that transfer readily from one local or regional ecosystem to another. One Health will be far more interested than the social–ecological systems theorists (so far) in the role of animals in people's lives (as livestock and companion animals, or wild species they eat or interact with), and in people's behaviour patterns within those interactions (e.g. animal tending practices, supply chain behaviour cf. Dudley 2008) that foster good health or risk disease spread.

One Health will thus be interested in the linkages between specific ecosystems, or at least in the ecological basis of particular pathogens and disease vectors that might readily (and often rapidly) find opportunities elsewhere. Given the rapidity of international travel, the transfers of animals and foods under a globalised economy, and influences of migratory bird movements (Cumming 2010), One Health should be interested in one global system made up of many localised, but also highly networked, sub-systems. One Health may find the concept of resilience very useful: since we cannot prevent diseases (of animals, humans or both) altogether, or control them entirely, how can we make vulnerable people and regions more resilient to their occurrence? How can adaptive capacity be understood and enhanced, towards achieving that resilience? What are the key components of adaptive capacity and resilience (Berkes and Ross 2013; Magis 2010; Armitage et al. 2011) in a One Health practice context? Since management of risk and control of diseases is beyond any single authority's control (human health or

Table 1 Paradigms that could inform social and cultural dimensions of One Health systems approaches

Paradigm	Key concepts, principles or tenets	Relevance to One Health, examples of applications	Key journals
Social–ecological systems	Takes a complex adaptive systems view of people–environment relations, with particular focus on the ecology. Contributes the idea of resilience. Focuses on interactions across multiple geographical, social and governance scales	Encourages a dynamic view of the relationships inherent in epidemiology, interpreted in terms of complexity. How do pathogens and vectors behave in changing contexts? How do multiple factors interact to produce outcomes and trends? How do humans, animals and ecosystems adapt to changing circumstances, including disturbances?	
Human ecology	Examines people–environment relationships comprehensively, with considerable emphasis on adaptation. It is particularly holistic when combined with anthropological approaches (ethnography). The early work tended to over-emphasise the relevance of ecological change processes (adaptation) towards human ones; later work has been comprehensive in theoretical development and case studies	The body of published work, especially ethnographies that include relationships with animals Adaptation concept, processes of adaptation	*Human Ecology* *Journal of Human Ecology* *Human Organization*
Political ecology	An offshoot of human ecology and political economy that emphasises power relationships in shaping environmental and social justice outcomes	Important for the recognition of how power relationships within society and governance can easily disadvantage the least powerful in society. Governance processes and landscape outcomes often favour the rich and powerful. In One Health, there is a risk of the least powerful in societies bearing the burden of disease control (e.g. in culling of animals)	*Political Ecology*
Environment–behaviour studies/environmental psychology	A field developed with respect to the built environment, usually at small scales e.g. neighbourhoods, housing. Studies the two-way interactions between people and their environments. Brings useful focus on cognitions, and concepts such as behaviour settings and activity systems	Available for elaboration to include human relationships with animals in natural and built settings. Encourages attention to built environments, such as how built forms and the behaviours within them mediate exposure to animals and disease risks	*Environment and Behaviour* *Journal of Environmental Psychology*

veterinary), what possibilities lie in looking towards adaptive governance ideas (Stirling and Scoones 2009), focused on collaboration across parties with capacity to address parts of a problem, and on taking adaptive rather than rigid approaches to solution finding.

Further, how can One Health manage the risks of conflicting activity between scales (levels within a system) when addressing crises? Adger et al. (2011) noted with respect to climate change adaptation that some case study countries worked in harmony with and empowered local people's adaptation, whereas others over-rode good local initiative and practice with imposed national policies in ways that damaged overall capacity. The analogies for One Health are obvious: how can local practice and arrangements be harmonised with national and international effort, rather than running the risk of contradictions, conflicts, inefficiencies and increased disease risk?

2.2 Important System Interactions in One Health

One Health is interested in interactions between humans, animals and ecosystems, and hence their systems of management, towards enhancing well-being, reducing risks and making management more effective. One Health will thus have particular foci of interest within the complexity of our global social–ecological systems (which includes economic systems). Human–animal–environment interactions of particular interest to One Health are:

- The many ways in which humans interact with animals, from the tending of livestock for livelihoods and also cultural reasons (where livestock may represent cultural status, savings, bridge-price or dowries), to interactions with wildlife for sustenance (hunting) and enjoyment, or accidental contact in the course of other practices, to the psychological benefits of human–companion animal interaction.
- Causes of different patterns of interaction between wildlife, domesticated species and humans—for instance, where seasons or the effects of environmental change (including climate change) affect species movements and disease vectors, or where loss of wildlife habitat through land clearing to meet population increases and economic demands leads to new interactions among different animals, and animals and humans.
- The effects of global movements of people and animals, for instance, in travel and trade (including supply chains, Dudley 2008).
- The policy and administrative interactions involved in issues that transcend (or force collision between) established systems of governance. (Western-influenced forms of governance typically keep health and environment separate, animal and human health separate and national and local government powers separate, where One Health requires strong networking and capacity for coordinated action across all of these).

3 Social and Cultural Dimensions Important to One Health

From the wealth of social science concepts available for understanding human thinking and behaviour, some key concepts appear particularly relevant to the points above. All of these feature within the other paradigms (see Table 1) which could well enrich missing dimensions in social–ecological systems. This analysis makes no claim to comprehensiveness: other social science concepts are potentially relevant. It concentrates on cultural, cognitive and behavioural dimensions of people–animal–ecosystem/environment interactions. Since a culture entails a coherent combination of many of these concepts, aspects of this analysis will overlap.

3.1 Cultural Dimensions

Cultures are known to develop, over lengthy periods, in relation to ecosystems and climate. Thus, hunter–gather societies, in general, are known to have some characteristics in common that differ markedly from those of agricultural societies. McMichael (2004) points out that major cultural shifts have also been associated with shifts in infectious disease threats. Cultures are commonly studied in terms of religion and belief systems, kinship, resource use, economic behaviour, development and use of technologies and built environment. Together, these (and other) characteristics shape interactions between people and their ecosystems, the people within the society and people and materials. Culture includes the social rules for relationships among people, and the nature of the supernatural worlds they create. Myriad permutations are possible, for instance, the diverse forms and meanings of vernacular housing and settlements developed by the world's different cultures (Rapoport 1969).

The relevance for One Health is extensive. Culture shapes (or represents) all of the social attributes explored below, and is a major part of defining—or reflecting—the ways in which humans, animals and ecosystems interact for economic and socially-defined purposes. It influences relationships among members of a society (e.g. gendered relationships), and the way communication occurs (e.g. Hickler 2007), and hence the strategies that One Health practitioners might need to adopt in addressing disease risks.

3.2 Cognitive Dimensions

This represents the set of individual and shared mental domains that are potentially relevant in people–animal–ecosystem interactions. Values express broad preferences concerning appropriate courses of action or outcomes, representing a person's or society's sense of right and wrong or desirable conditions. In a One Health context, we may be interested in values towards particular species, specific environments and practices (whether culturally or socially endorsed, or discouraged). Values are closely related to social and behavioural norms; the social 'rules' for behaviour in a society are taught and enforced through social influence. Norms tend to be more transient and mutable than values; for instance, norms for food handling have changed as a matter of public education and social influence.

While the term 'attitudes' (orientation or aversion towards a certain type of object) is commonly (and often erroneously) used to refer to a broad suite of cognitive dimensions, other concepts such as 'mental models' (Jones et al. 2011) are likely to be more useful for understanding the system inherent in One Health. Mental models are individual and collective understandings of how a system works. They tend to focus on cause-effect relationships (real or believed), but may include emotional aspects such as a positive or negative orientation towards an object (such as a species of animal, of a health promotion practice) within the system. A mental model's focus on One Health would lead practitioners to explore how a society and individuals within it understand particular disease patterns to occur and also missing assumptions; for instance, absence of recognition (or even denial) of risks that may be well-known to epidemiologists. Mental models can incorporate beliefs (holding a proposition to be true) and knowledge (theoretical or practical understanding of a subject). For a One Health context, knowledge is best considered comprehensively, as including a culture or individual's world view. Many societies include spiritual dimensions within their cosmologies, and people–animal–ecosystem relationships may feature here.

Various other social science concepts are potentially relevant. The geography and psychology concept of 'sense of place' links the cognitive to an ecosystem or built environment, to express both characteristics of the place and the beholders' sense of identification with that place (for instance, a sense of belonging to and familiarity with, love of, a locality). In a One Health context, sense of place may contribute to public support for keeping ecosystems healthy, or underpin resistance to making change which may be necessary from a risk prevention perspective. It would be valuable to elaborate the idea of 'sense of place' to 'sense of relationship with species'. For instance, in the author's region of South East Queensland, Australia, the Hendra virus which transmits from flying foxes, a protected species, to horses and thence to humans (Field et al. 2007) has brought out social conflict between horse-owners and wildlife proponents, each identifying strongly with one species and at worst seeking the removal of the other.

'Identity' also appears very important in people–animal–ecosystem relationships. This concept spans an individual sense of self to social and cultural groupings' self-attribution of characteristics. The relevance for One Health is that needs for disease control can conceivably cut across peoples' strongly held identities. For instance, herding societies may be strongly reticent to have their animals culled, obviously for livelihood reasons but also because of strong cultural identification as 'herding peoples', and most likely, with social status being associated with herd strength. In remote Aboriginal Australia, dogs have an ambivalent role. They are little-tended and may harbour a variety of pathogens, yet their cultural meaning is such that while they may be neglected under (Kimberley) Aboriginal law, they may not be killed (Ross 1987).

Another concept worth borrowing is that of 'meaning', used very effectively in relation to the built environment by Rapoport (1969). What is the meaning of particular species in particular cultures? Why is that species important, e.g. livestock have different meanings to the Masai and other East African herders than to Southeast Asian mixed-subsistence farmers.

A variety of other cognitive dimensions could also be considered, for instance, social constructions of the relationship between people and animals (after Greider and Garkovich 1994).

3.3 Behavioural Dimensions

Having considered the options for understanding peoples' mental interactions with animals and ecosystems, we need to consider their behaviour patterns. Since One Health is interested in managing key disease transmission patterns, key questions are:

- What do people do with animals in their built, farmed and natural environments, and why?

Common reasons for interactions and interdependencies among people, animals and ecosystems are livelihoods (through hunting and gathering, for food or medicinal purposes; and farming, even bridge-price and dowries where animals are essential to economic exchanges on marriage, and marriage is a matter of livelihood as well as inter-familial diplomacy). Further reasons include companionship, in domestic animals, and aesthetic appreciation (as in wildlife tourism, and enjoyment of animals in the wild).

Behaviour patterns of interest will include how safely people actually interact with the animals and related environmental contaminants. Are animal husbandry practices reducing or exacerbating the risks of disease transmission? Then, how amenable are these practices to change? Are they deeply embedded in cultural practices, or more peripheral and thus less likely to provoke resistance should change be advocated?

- What is the role of animals (wild and domesticated) in economic relationships?

Animals play important roles in the livelihoods of many societies. Common roles for animals are when livestock is tended as a regular source of food or cash income (e.g. herding and grazing societies such as Australian commercial graziers, migratory herders in Eastern Africa such as Somali and Masai); when animals are tended but retained as a form of banking and often social status, as part of mixed farming systems (e.g. pigs and poultry can represent savings, for later sale, or dowries; in such societies they often represent social status); livestock as working animals (e.g. draught horses and buffalo). There is a set of subsistence roles in harvesting of wild species directly for consumption as food or medicine, but sometimes also for intermittent sale. This is common in forest ecosystems.

- Does the role of the environment in livelihoods, e.g. land clearing for more agriculture, press wild animals into greater interaction with domesticated species and with increased risk of zoonoses?
- What are the roles of animals in companionship and domestic settings, and what health risks and benefits attach to these?

A useful concept from environmental psychology is that of behaviour settings (Wicker 1972). While the initial theory focused strongly on built environments, it is worth extending to such settings as fields and stables or livestock pens, and to animal as well as human behaviour. The concept deals with certain practices occurring in certain settings: the practices are not caused (although influenced) by the physical features of the setting; they are strongly influenced by social norms about what should occur where. For example, availability of hand washing facilities near a livestock pen may facilitate cleanliness after handling animals, but the behaviour of hand washing needs to be a social and individual norm, otherwise it is unlikely to occur.

3.4 Other Processes

A range of other processes involving interactions between people also deserve a part in a socially and culturally aware One Health. Education, and particularly the related social process of social learning, is important to benefit from experience and secure improvements to interactive patterns that threaten human, animal and ecosystem health. Social networks, and the social influence that commonly occurs through such relationships, appear important in achieving practice change and distributing learning. This may include acceptance of necessary disease control strategies. Communication processes are highly important in disease control, both through community-based interactions and through formal processes. The form and success of communication relates closely to social and cultural factors (Hickler 2007).

4 Research Directions

This analysis of social and cultural dimensions worthy of exploration to assist the development of One Health suggests a set of research issues to guide social scientists and interdisciplinary teams. First, what is the nature of the interactions among people, animals and environment in a specified context? What role and meaning do certain animal species have in this culture? For example do they have roles in livelihoods, companionship and important cultural symbolism? Ethnographic analyses of this type might include avoided interactions, such as under taboos, and spiritual and stewardship relationships, as with animal totems. What are the economic dimensions in these interactions, e.g. subsistence, livelihoods including treating certain animals as savings for later sales or dowries. Do these relationships occur in ways conducive to health risks (disease risks or risk of spread) or produce benefits (e.g. interaction such as animal therapies)? Do they present significant issues for disease control, for instance, in reluctance to part with favoured animals during culls?

Second, how should risk mitigation and crisis handling be conducted in these cultural contexts? How do culture, values, behaviour patterns and economic dependencies affect choices of action pathways—such as communication, education, actions, and compensation for animals destroyed? Given the difficulties of conducting ethnographic studies when risks emerge, can participatory processes help to pool knowledge and understandings, and identify socially and culturally acceptable solutions? How can community and public 'engagement' be conducted in One Health contexts, to achieve ownership and involvement on the part of the community (Aslin and Brown 2004 p. 5)?

How can forms of governance in health promotion and disease control processes navigate both the complex adaptive systems inherent in people–animal–ecosystem health and disease relationships? What can be learnt from collaborative and adaptive forms of governance, as espoused in the social–ecological systems literature? How can the more familiar 'top-down' (directive) approaches used particularly in emergencies draw upon 'bottom up' approaches that might inform more culturally sensitive and socially acceptable—and hence workable—solutions?

5 Conclusions

System understanding and system strengthening for One Health would benefit from use and elaboration of bodies of theory that deal with people–environment relationships. All of these need elaboration to incorporate animal relationships more explicitly. This chapter advocates use of the social–ecological systems body of theory for its use of the complex adaptive systems paradigm and understanding of dynamic forms of behaviour, but notes worthwhile concepts from human ecology, political ecology and environment behaviour studies.

Within such systems understandings, One Health needs to incorporate social and cultural considerations well. This is necessary to achieve a rich picture systems understanding and develop sensitive and effective systems strengthening to enable responsiveness in crises and slower moving change processes, and address risks and opportunities alike. Doing so involves considerable challenge, given the wide variety of social and cultural issues that are potentially relevant, and the wide variety of social science (and humanities) disciplines available to contribute. We do not want the situation of 'hammers finding nails'; single disciplinary approaches or working on single concepts at a time would not contribute to effective systems approaches. We thus would do better to incorporate exploration of social and cultural dimensions within systems frameworks that consider human–environment relationships, and expand these to explore the particular linkages that apply around One Health research and practice questions. This entails more explicit recognition of the roles of animals, and translation of the older frameworks to recognise the 'complexity' paradigm of complex adaptive systems.

Acknowledgments The author thanks collaborators in a recent study of zooneses, for introducing her to the work in One Health and expanding her knowledge of zoonoses. She particularly acknowledges the influences of Helen Scottorr, John Mackenzie, Lisa Adams, Jane Goller, Ian Patrick and Anne Ancia.

References

Adger WN, Brown K, Nelson DR, Berkes F, Eakin H, Folke C, Galvin K, Gunderson L, Goulden M, O'Brien K, Ruitenbeek J and Tompkins EL (2011) Resilience implications of policy responses to climate change. WIREs Clim Change, 2:757–766. doi:10.1002/wcc.133

Armitage D, Berkes F, Dale A, Kocho-Schellenberg E, Patton E (2011) Co-management and the co-production of knowledge: learning to adapt in Canada's Arctic. Glob Environ Chang 21:995–1004

Asia Pacific Economic Cooperation (2011) APEC One Health action plan: a framework to assist APEC economies strengthen cross-sectoral networks and functioning against the threat of emerging and zoonotic infectious diseases. http://www.dld.go.th/dcontrol/th/images/stories/OneHealth_2011/doc-onehealth/apec_one_health_action_plan_september_2011.pdf. Accessed 30 Aug 2012

Aslin H, Brown V (2004) Towards whole of community engagement: a practical toolkit. Murray-Darling Basin Commission, Canberra

Berkes F, Folke C (eds) (1998) Linking social and ecological systems: management practices and social mechanisms for building resilience. Cambridge University Press, Cambridge

Berkes F, Ross H (2013) Community resilience: towards an integrated approach. Soc Nat Resour 26

Berkes F, Folke C, Colding J (eds) (2003) Navigating social–ecological systems: building resilience for complexity and change. Cambridge University Press, Cambridge

Caron A, de Garine-Wichatitsky M, Gaidet N, Chiweshe N, Cumming GS (2010) Estimating dynamic risk factors for pathogen transmission using community-level bird census data at the wildlife/domestic interface. Ecol Soc 15:25. URL:http: //www.ecologyandsociety.org/vol15/iss3/art25/ [online]

Cote M, Nightingale AJ (2012) Resilience thinking meets social theory: situating social change in socio-ecological systems (SES) research. Prog. Hum. Geogr. 36:475–489

Cumming GS (2010) Risk mapping for avian influenza: a social–ecological problem. Ecol Soc 15:32. URL: http://www.ecologyandsociety.org/vol15/iss3/art32/ [online]

Davidson DJ (2010) The applicability of the concept of resilience to social systems: some sources of optimism and nagging doubts. Soc Nat Resour 23:1135–1149

Dudley, JP (2008). Public health and epidemiological considerations for avian influenza risk mapping and risk assessment. Ecol Soc 13:2. URL:http://www.ecologyandsociety.org/vol13/iss2/art21/ [online]

FAO, OIE, WHO, UN System Influenza Coordination, The World Bank (2008) Contributing to One World, One Health: a strategic framework for reducing risks of infectious diseases at the animal–human–ecosystems interface. http://www.fao.org/docrep/011/aj137e/aj137e00.htm. Accessed 30 Aug 2012

Field HE, Mackenzie JS, Daszak P (2007) Henipaviruses: emerging Paramyxoviruses associated with fruit bats. CTMI 315:133–159

Folke C (2006) Resilience: the emergence of a perspective for social–ecological systems analyses. Glob Environ Chang 16:253–267

Greider T, Garkovich L (1994) Landscapes: the social construction of nature and the environment. Rural Sociol 59:1–24

Gunderson L, Holling C (eds) (2002) Panarchy: understanding transformations in systems of humans and nature. Island Press, Washington

Gunderson L, Holling C, Light S (eds) (1995) Barriers and bridges to the renewal of ecosystems and institutions. Columbia University Press, New York

Hickler B (2007) Bridging the awareness between HPAI 'awareness' and practice in Cambodia. Emergency Centre for Transboundary Animal Disease, FAO Regional Office for Asia and the Pacific

Jones NA, Ross H, et al (2011) Mental models: an interdisciplinary synthesis of theory and methods. Ecol Soc 16:46. URL:http://www.ecologyandsociety.org/vol16/iss1/art46/ [online]

Magis K (2010) Community resilience: an indicator of social sustainability. Soc Nat Resour 23:401–416

McMichael AJ (2004) Environmental and social influences on emerging infectious diseases: past, present and future. Phil Trans R Soc Lond B 359:1049–1058. doi:10.1098/rstb.2004.1480

Olsson P, Gunderson L, Carpenter S, Ryan P, Lebel L, Folke C, Holling C (2006) Shooting the rapids: navigating transitions to adaptive governance of social–ecological systems. Ecol Soc 11:18

Olsson P, Folke C, Galaz V, Hahn T, Schultz L (2007) Enhancing the fit through adaptive co-management: creating and maintaining bridging functions for matching scales in the Kristianstads Vattenrike Biosphere Reserve, Sweden. Ecol Soc 12:28

Rapoport A (1969) House form and culture. Prentice Hall, EnglewoodCliffs, New Jersey USA

Ross H (1987) Just for living: Aboriginal perceptions of housing in Northwest Australia. Aboriginal Studies Press, Canberra

Ross H, Poungsomlee A, Punpuing S, Archavanitkul K (2000) Integrative analysis of city systems: Bangkok Man and the Biosphere Programme study. Environ Urban 12:151–161

Si Y, Wang T, Skidmore AK, de Boer WF, Li L, Prins HHT (2010) Environmental factors influencing the spread of the highly pathogenic avian influenza H5N1 virus in wild birds in Europe. Eco Soc 15:26. URL: http://www.ecologyandsociety.org/vol15/iss3/art26/ [online]

Stirling AC, Scoones I (2009) From risk assessment to knowledge mapping: science, precaution and participation in disease ecology. Eco Soc 14:14. URL: http://www.ecologyandsociety.org/vol14/iss2/art14/ [online]

Walker B, Salt D (2006) Resilience thinking: sustaining ecosystem and people in a changing world. Island Press, Washington

Walker B, Westley F (2011) Perspectives on resilience to disasters across sectors and cultures. Eco Soc 16:4. URL: http://www.ecologyandsociety.org/vol16/iss2/art4/[online]

Walker B, Holling C, Carpenter S, Kinzig A (2004) Resilience, adaptability and transformability in social–ecological systems. Eco Soc 9:5 [online]

Westley F, Carpenter S, Brock W, Holling C, Gunderson L (2002) Why systems of people and nature are not just social and ecological systems. In: Gunderson L, Holling C (eds) Panarchy: understanding transformations in systems of humans and nature. Island Press, Washington

Wicker AW (1972) Processes that mediate behavior-environment congruence. Behav Sci 17:265–277

Wilson G (2012) Community resilience and environmental transitions. Earthscan, Oxford

Index

Current Topics in Microbiology and Immunology (2013) 366: 231–235
DOI: 10.1007/978-3-642-35846-3

Z

MIX
Papier aus verantwortungsvollen Quellen
Paper from responsible sources
FSC® C105338

If you have any concerns about our products,
you can contact us on
ProductSafety@springernature.com

In case Publisher is established outside the EU,
the EU authorized representative is:
Springer Nature Customer Service Center GmbH
Europaplatz 3, 69115 Heidelberg, Germany

Printed by Libri Plureos GmbH
in Hamburg, Germany